MULTIVARIATE DATA ANALYSIS IN SENSORY AND CONSUMER SCIENCE

F N P

PUBLICATIONS IN FOOD SCIENCE AND NUTRITION

Books

MULTIVARIATE DATA ANALYSIS, G.B. Dijksterhuis
NUTRACEUTICALS: DESIGNER FOODS III, P.A. Lachance
DESCRIPTIVE SENSORY ANALYSIS IN PRACTICE, M.C. Gacula, Jr.
APPETITE FOR LIFE: AN AUTOBIOGRAPHY, S.A. Goldblith
HACCP: MICROBIOLOGICAL SAFETY OF MEAT, J.J. Sheridan *et al.*
OF MICROBES AND MOLECULES: FOOD TECHNOLOGY AT M.I.T., S.A. Goldblith
MEAT PRESERVATION, R.G. Cassens
S.C. PRESCOTT, PIONEER FOOD TECHNOLOGIST, S.A. Goldblith
FOOD CONCEPTS AND PRODUCTS: JUST-IN-TIME DEVELOPMENT, H.R. Moskowitz
MICROWAVE FOODS: NEW PRODUCT DEVELOPMENT, R.V. Decareau
DESIGN AND ANALYSIS OF SENSORY OPTIMIZATION, M.C. Gacula, Jr.
NUTRIENT ADDITIONS TO FOOD, J.C. Bauernfeind and P.A. Lachance
NITRITE-CURED MEAT, R.G. Cassens
POTENTIAL FOR NUTRITIONAL MODULATION OF AGING, D.K. Ingram *et al.*
CONTROLLED/MODIFIED ATMOSPHERE/VACUUM PACKAGING, A.L. Brody
NUTRITIONAL STATUS ASSESSMENT OF THE INDIVIDUAL, G.E. Livingston
QUALITY ASSURANCE OF FOODS, J.E. Stauffer
SCIENCE OF MEAT & MEAT PRODUCTS, 3RD ED., J.F. Price and B.S. Schweigert
HANDBOOK OF FOOD COLORANT PATENTS, F.J. Francis
ROLE OF CHEMISTRY IN PROCESSED FOODS, O.R. Fennema *et al.*
NEW DIRECTIONS FOR PRODUCT TESTING OF FOODS, H.R. Moskowitz
ENVIRONMENTAL ASPECTS OF CANCER: ROLE OF FOODS, E.L. Wynder *et al.*
PRODUCT DEVELOPMENT & DIETARY GUIDELINES, G.E. Livingston, *et al.*
SHELF-LIFE DATING OF FOODS, T.P. Labuza
ANTINUTRIENTS AND NATURAL TOXICANTS IN FOOD, R.L. Ory
UTILIZATION OF PROTEIN RESOURCES, D.W. Stanley *et al.*
POSTHARVEST BIOLOGY AND BIOTECHNOLOGY, H.O. Hultin and M. Milner

Journals

JOURNAL OF FOOD LIPIDS, F. Shahidi
JOURNAL OF RAPID METHODS AND AUTOMATION IN MICROBIOLOGY, D.Y.C. Fung and M.C. Goldschmidt
JOURNAL OF MUSCLE FOODS, N.G. Marriott, G.J. Flick, Jr. and J.R. Claus
JOURNAL OF SENSORY STUDIES, M.C. Gacula, Jr.
JOURNAL OF FOODSERVICE SYSTEMS, C.A. Sawyer
JOURNAL OF FOOD BIOCHEMISTRY, N.F. Haard, H. Swaisgood and B. Wasserman
JOURNAL OF FOOD PROCESS ENGINEERING, D.R. Heldman and R.P. Singh
JOURNAL OF FOOD PROCESSING AND PRESERVATION, D.B. Lund
JOURNAL OF FOOD QUALITY, J.J. Powers
JOURNAL OF FOOD SAFETY, T.J. Montville and D.G. Hoover
JOURNAL OF TEXTURE STUDIES, M.C. Bourne and M.A. Rao

Newsletters

MICROWAVES AND FOOD, R.V. Decareau
FOOD INDUSTRY REPORT, G.C. Melson
FOOD, NUTRITION AND HEALTH, P.A. Lachance and M.C. Fisher

MULTIVARIATE DATA ANALYSIS IN SENSORY AND CONSUMER SCIENCE

by

Garmt B. Dijksterhuis, Ph.D.

ID-DLO, Institute for Animal Science and Health
Food Science Department
Lelystad
The Netherlands

FOOD & NUTRITION PRESS, INC.
TRUMBULL, CONNECTICUT 06611 USA

FOOD & NUTRITION PRESS, INC.
6527 Main Street, POB 374
Trumbull, Connecticut 06611 USA

Library of Congress Catalog Card Number: 97-061692
ISBN: 0-917678-41-9

DEDICATION

To my father and mother

PROLOGUE AND ACKNOWLEDGMENTS

This book is the result of research into the applicability of Multivariate Data Analysis to the results of sensory studies. During the years I worked on this topic, I learned a lot, and had the opportunity to write down some of the things I had just learned. Of course, the credits are not all mine: I owe a lot to my teachers and colleagues, some of which appear as first or co-authors of papers in this book. I want to spend some words thanking them while at the same time sketching the history of this book.

Near the finishing of my study in psychology with Prof. Ep Köster at the University of Utrecht, Stef van Buuren suggested Overals as an interesting alternative for Procrustes Analysis, to analyse sensory data. This set me off in the direction of what I would now call Sensometrics. In 1987 I started working at Oliemans Punter & Partners, a small company that performs sensory and consumer research. The cooperation with Pieter Punter resulted, among other things, in a joint paper on Procrustes Analysis. Pieter would always put my nose in the direction of the applicability of MVA for sensory problems, which were useful lessons for me. In retrospect it occurs to me that I wrote almost all papers while I worked there. This is quite uncommon for such a small private company. I'm afraid I never explicitly thanked them for this, but hope to have put it right now.

The cooperation with Eeke van der Burg resulted in a number of papers, four of which are included in this book. I learned a lot from our cooperation, especially about the Gifi system and in particular about canonical analysis, redundancy analysis and their nonlinear extensions. Eeke is the first author of these four papers, which shows in the mathematical introductions. I thank her for never becoming tired when over and over again explaining some of the mathematics to me.

Another inspiring teacher was John Gower. His telling me about high-dimensional intersections of category-hyperplanes, with appropriate gesticulation and scribbles on the blackboard gave me another view on data analysis. We wrote two papers together of which one is included in the book. John is the first author, which shows in the generality of the method and its mathematical presentation.

In addition to teachers I thank my former colleagues at OP&P's for the discussions about a gamut of topics, some of which were sensory science and statistics. Margo Flipsen and Els van den Broek deserve special mention. They visited OP&P to do some Time-Intensity studies for their master's thesis at the Agricultural University of Wageningen. They appear as co-authors on two papers on the analysis of TI-data.

This book served as my Ph.D. thesis, at the department of Datatheory, at the University of Leiden. The main threat to the thesis ever coming to an end was I. Every now and then I would lose myself in a "very interesting" side-track of Multivariate Data Analysis. It was Willem Heiser who, by patiently and repeatedly telling me that I should focus on "sensory applications", put me back on the track again. Over the years he must have told me this several times, and it helped.

Ann Noble (University of Davis, California, USA) had become a kind of e-mail consultant to me. I thank her for her prompt answering of my questions, providing references, and commenting on some of my writing.

My current job is at the Food Science Department of ID-DLO, the Institute for Animal Science and Health (Lelystad, the Netherlands), leading their sensory laboratory. ID-DLO is one of the major research institutes on animal production. In their Food Science Department resides the research on the eating quality and safety of meat, eggs and dairy products mainly, in relation to the processing required to produce a palatable food. At this sensory laboratory I plan to explore some of the newer directions in sensory and consumer science outlined in this book.

Finally there are a number of people that, in some way or another, helped with the finishing of this book. To be sure to include them all, I do not give names, but I thank them all. However, one name must be mentioned. Because the preparation of the thesis was not part of my job, a lot of the writing took place at home, Gerjo is thanked for her patience, enthusiasm and organisational talents I needed to finish this project.

GARMT B. DIJKSTERHUIS
AMMERSTOL

CONTENTS

PART I: INDIVIDUAL DIFFERENCES

PART II: MEASUREMENT LEVELS

PART III: SENSORY-INSTRUMENTAL RELATIONS

PART IV: TIME-INTENSITY DATA ANALYSIS

CHAPTER 1

Introduction

Summary

In this introduction the basic terminology of the subject matter of this book is introduced. The underlying research question of the study is presented and the four main themes that the chapters cover are described and related to this question. The kind of data that are analysed throughout the book is explained. Furthermore, the aim and the structure of the book are explained.

1.1 Research Question

This book is concerned with problems from sensory and consumer research. What exactly these kinds of research are is defined later. The tools used to study the problems are the apparatus of Multivariate Data Analysis. The underlying question that is addressed by the research in this book is:

> What has Multivariate Data Analysis to offer for studying problems in sensory research?

Asking a question like this is more or less "upside down". Usually when one is confronted with a scientific problem, an experiment is designed to study it or to test a hypothesis. The reason that this is turned around in this book is that the problems in sensory science that the author was confronted with, were often such that the data were already collected. It was the feeling of both the sensory researchers and the author that "There's more than meets the eye in this data set."

1.2 Sensory Science

Sensory science is the general heading under which the study of many different problems and the application of methods can be found. No complete picture of sensory science will attempted to be given here. A concise history and overview of the field can be found in Stone and Sidel (1985, 1993), McBride (1990) and Punter (1991).

1.2.1 Some Definitions

The part of Sensory Science that this book is concerned with can be defined as follows:

> Sensory evaluation is a scientific discipline used to evoke, measure, analyse and interpret reactions to those characteristics of foods and materials as they are perceived by the senses of sight, smell, taste, touch and hearing.

This definition was used by the (U.S.) Institute of Food Technologists in 1975 and quoted by Stone and Sidel (1985). The definition is very general, but it contains most ingredients of the discipline as it will be presented in this book. The focus in this book will be on the *analysis* of the *reactions* to certain *characteristics* of *food* products (italics refer to the ingredients of the definition). The *reactions* to *characteristics* will be in the form of scores given to attributes perceived in the *food*-stimuli, the analyses will be multivariate and the *senses* will mainly be the senses of *smell* and *taste*.

The field has many names, which may stress different aspects of Sensory Science, but globally the same problems underlie all sub-disciplines. Thomson (1988) poses the question:

> What are the attributes that consumers perceive in a particular new food product and in what ways will these combine to determine future purchase decisions?

as one of the most obvious questions to be answered by the scientific discipline coined "Food Acceptability". He also describes "Food Acceptability" as a somewhat uncomfortable marriage between food science and behavioural psychology. In an attempt to consolidate this marriage, a third party is introduced in this book: data analysis.

McBride (1990) gives an overview of the position of sensory evaluation in between the other disciplines:

1. research and development with a food-technical focus
2. consumer and marketing research with a behavioural and psychological focus

Note that the marriage Thomson (1988) referred to is reflected here too. A lot of bridges can be, and are being, built between the different disciplines involved (see e.g. Thomson *et al.* 1988). In this book a bridge is being developed based on statistics and data analysis.

1.2.2 Sensory and Consumer Science and Related Disciplines

A brief layout along simple lines will be given here to explain further the subject matter of this book. From now on the term *Sensory and Consumer Science* will be adopted, because it reflects reasonably well the contents of the field. It is set apart from the study of the chemical senses, which is commonly referred to as Sensory Psychology and Sensory Physiology (e.g. Köster 1971, de Wijk 1989). Such research is not of concern in this book. The (chemical) senses can also be studied in connection to psychological properties of the experimental subject. In this case behavioural responses may (be attempted to) be modelled mathematically and the properties of the models studied. This kind of research is historically linked to psychophysics and psychometrics. Recent psychophysical studies with applications in sensory science and psychophysics were performed by Frijters (1980) and Ennis (1991). This field is again not the subject of this book.

Figure 1[1] presents an overview of the different parts *Sensory and Consumer Science* contains.

---

[1]This figure is based on a suggestion by Pieter Punter.

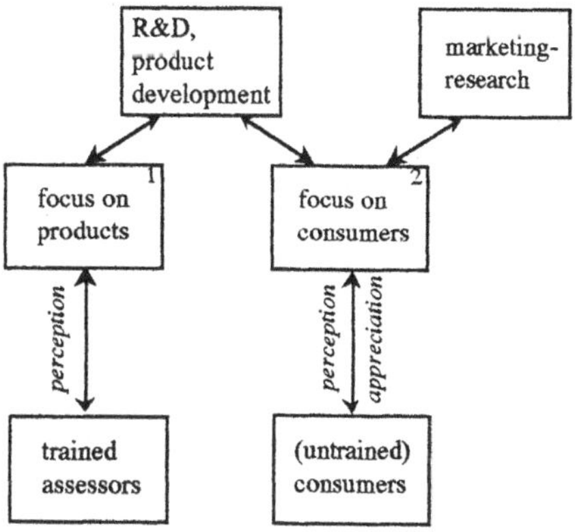

Figure 1 Overview of Sensory and Consumer Science, illustrating the differences in focus (1: on products; 2: on consumers).

As is illustrated in Figure 1, Sensory and Consumer Science and marketing/consumer research can be subdivided into two main fields:

1. the study of products
2. the study of consumers

In the study of products, mainly trained assessors are used to judge the products on rather technical or analytical attributes. This is what is meant by *perception*, in the figure, in contrast to *appreciation*. The hedonic quality *appreciation* of the products is of no concern in this type of sensory research.

Both appreciative and perceptive aspects are used in the consumer focused studies. The perceptive part uses consumer characterisations of the products, rather than technical/analytical attributes. The appreciative part may include measurement of the ideal intensity of the attributes and/or the preferences of the consumers.

Product-oriented research has a clear relation to R&D and product development. Consumer-oriented sensory research in addition has a relation to marketing research.

In this book, the focus is on the products rather than on the consumer. The perception will mainly involve taste and smell properties of the products, though visual, auditory (e.g. Vickers 1983, 1991) and kinaesthetic perceptions are by no means excluded from Sensory and Consumer Science.

The distinction in Figure 1 is not so strict as the figure may suggest. Sensory profiling studies are usually of an analytical nature, hence often found in perception-studies. They try to answer the question: "What are the important attributes of the products?" They can be applied in appreciation studies too. Then the question: "What products are preferred/accepted/appreciated by the consumers?" is answered. It is a matter of choosing the attributes. In appreciation studies, the attributes are fixed and will be mostly hedonic and focusing on aspects of the quality of the products. Profiling studies will be introduced in more detail in later sections (§1.4).

In Sensory-Instrumental research, the relations between physical/ chemical (*instrumental*) properties and the sensory properties of products are studied. The focus of these studies is mainly analytical, i.e., they are perception-studies. However, they may be conducted in an appreciation context, provided that special attention is given to the relations between the instrumental and the sensory-appreciative (see e.g. Noble 1975). Sensory-Instrumental research is covered in more detail in section 1.7 and is the subject of Part III.

Time-Intensity research (§1.8 and Part IV) is focused on perception only. The time-course of a particular perceived property of a product is studied.

1.3 Sensory Research and Sensory Profiling Data

The questions dealt with in this book are from the field of sensory and consumer science. In general terms, this is the field of research in which *people* use their senses to *describe* certain *properties* of *objects*. Admittedly this definition is too general and needs narrowing.

Three entities constitute the research and the resulting data in this book:

- Objects
- People
- Descriptions (of properties)

Objects can be interpreted very broadly. People can describe physical objects, other people, services, etc. Other terms used are *products* or, borrowed from psychology, *stimuli*.

The *descriptions* can take different forms. They can be a judgement of the quality of an object, its hedonic value or another specific property. In this book, the descriptions will take the form of judgements of a particular sensory property of the object, e.g. its sweet taste, its colour, its bitterness or the roughness of its surface. These properties will be called *attributes*, and they constitute the *variables* of the research in the sequel. A variable may consist of numerical scores, or of a number of (ordered) categories.

In sensory research the data are almost exclusively elicited from *people*. One of the directions in sensory and consumer science is research of products with the use of sensory panels, sensory profiling studies. A sensory panel is a group of people who give judgements about products. There are different kinds of sensory panels, some of which will be introduced in a following section.

The products in the case of sensory research are food products, drinks, cosmetics or luxuries like snacks, candy or tobacco. The products are evaluated using essentially all senses (sight, hearing, smell, touch and taste) though depending on the specific research question the focus may be on just one or two of them. In purely analytical taste and/or smell studies, the appearance of different products will be controlled for by e.g. using special lighting conditions. Another modality is texture perception in the mouth. This sense is important when judging products where texture plays a role e.g. in meat. Sight and even hearing also play a part in sensory research. The appearance of products may be important, depending on the kind of research. The sound of potato chips during chewing is an example of use of the auditive sense in judging edible goods (see also Vickers 1991).

1.3.1 Sensory-, Consumer- and Marketing Research

Sometimes the line between sensory, consumer and marketing research is very thin indeed. Often a sensory panel receives a certain amount of training in the judging task that is expected of them. The term *consumer panel* is sometimes reserved for a group of judges that are not trained with respect to their task. They are sometimes described as (or in fact) "picked up from the street", but it also happens that such a panel receives a limited amount of training. No clear standard terminology seems to exist. Matters may get more complicated when the term marketing-research is included in the picture too. Is sensory research a special case of consumer research, which is a special case of marketing research? It proves hard to answer this question and perhaps it is even harder to consolidate sensory researchers with consumer and marketing researchers. Van Trijp (1992, see also Figure 2) makes a distinction between the different types of product that are studied by the different disciplines. Sensory research studies the *core product*, i.e. a product with certain physical/chemical ("instrumental") characteristics of which the sensory characteristics are sought. This is the study of the *perception* of products as presented in §1.2.2. The *generic product* possesses certain derived "benefits" as usage utility, ease of use, perceived durability and a "status". This generic product is different from the core product, though the same physical product may underlie both. Consumer or marketing research is concerned with studying the generic products.

Figure 2 illustrates the relations between the fields of sensory, consumer and marketing research.

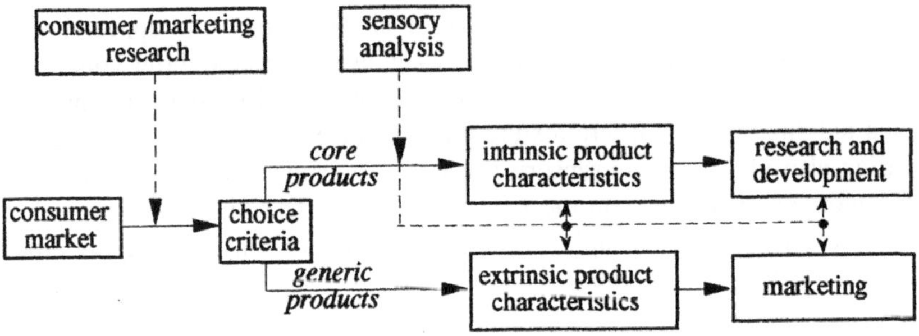

Figure 2 Relations between sensory, consumer and marketing research, showing the differences between *core products* and the *generic products* (slightly adapted from van Trijp 1992).

Figure 2 shows the "classic" point of impact of sensory analysis, studying the intrinsic product characteristics (the *core* product) for research and development. The two double arrows between the intrinsic and extrinsic product characteristics, and between R&D and marketing, indicate an interesting potential application of sensory analysis and marketing, viz. the study of to what extent sensory perception is influenced by properties of the "generic" product such as price, packaging, brand labelling, and the derived characteristics of the generic product.

Sensory research and consumer/marketing research have different, though both psychological, origins. Sensory research is based in sensory physiology and psychology and has, through psychophysics, always had a link with statistics and psychometrics (see e.g. Punter 1991). Marketing/consumer research has its origins in social psychology, and it has a strong link to direct applications in marketing. Sensory research is perhaps less applied than marketing/consumer research, in that it is closer to research and development of products, and further away from the market (see also §1.2.1, Thomson 1988, McBride 1990).

1.3.2 Sensory Panels and Ditto Data

There are a number of different ways to collect sensory profiling data, using different kinds of sensory panels. One important aspect in which these methods of data collection differ is in the amount of training of the panels receive prior to the actual experiment. Figure 3 arranges the different panel-types along a continuum with respect to the amount of training they receive.

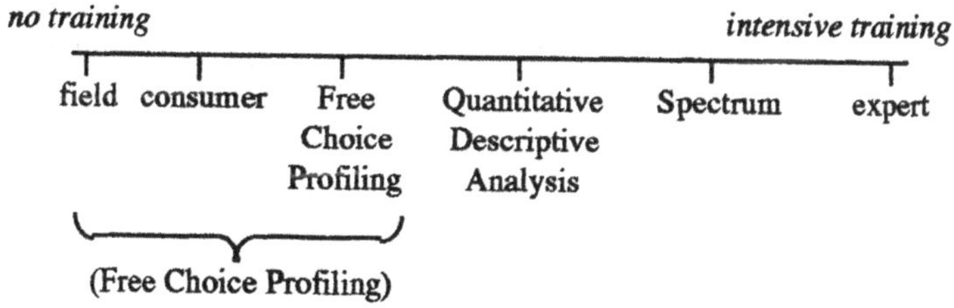

Figure 3 "Sensory panel method continuum", ranging from untrained panels at the left to panels that receive much training at the right.

The sensory analytical panels are located at the right extreme of this continuum. These panels judge a limited set of products on a number of strictly defined properties, with respect to which they have been intensively trained. They are sometimes referred to as *expert* panels. At the other end of the continuum in Figure 3 the consumer panels reside. Here one moves closer to marketing research. The most extreme example is probably found in "mobile testing" where the research takes place in a prepared bus which drives up to a shopping centre and invites people in to judge products. These panels may be called *field* panels, to distinguish them from *consumer* panels in which inexperienced consumers are invited to take place in a sensory experiment inside a laboratory, or at least in a somewhat more controlled environment than a bus. In between the field-panels and the expert-panels a lot of different sensory-panel methods exist of which some are indicated in Figure 3. The differences between the *QDA* panel and the *Spectrum* panel method are not fundamental, and they are not explained here (see Stone and Sidel 1985, 1993 for QDA, Meilgaard *et al.* 1990 for Spectrum). These two methods have in common that a standard vocabulary of descriptive attributes is formed. These attributes are used in the sensory experiment after the panel receives training with respect to the attributes. The differences between the methods lie in the amount of and procedures of training of the sensory panel.

Another way of distinguishing between different panels is in terms of the kind of questions the judges are asked. A sensory panel is also referred to as an *analytical* sensory panel when the questions in the experiment apply to analytical, as opposed to hedonic, properties of the products. This division is also present in Figure 1, where the term "perception studies" is used for "analytical," and "appreciation studies" for "hedonic studies." Examples of analytical attributes are sweet taste, nutty taste, sticky odour, rubbery texture, etc. The further we move to the right on the continuum in Figure 3, the less likely it is that hedonic questions will be asked. Hedonic studies are not explicitly covered in this book. However, when analytical attributes are replaced by hedonic attributes, or just by one hedonic attribute, most MVA methods discussed in this book can be used for hedonic sensory profiling studies as well.

Free Choice Profiling panels differ not only in the amount of training, but also in another property (see §1.4.2). This is why it is hard to include FCP panels in Figure 3. The panels that are usually called FCP panels are at the approximate position indicated in Figure 3. They often contain consumers, or somewhat more experienced panelists, who receive only a limited amount of training with respect to the attributes. The important property of FCP panel studies is that the assessors can choose their own attributes. When field or consumer panels are allowed to choose their own attributes they become FCP-panels too, hence the brace in Figure 3. The panels at the right hand side of FCP on the continuum are not FCP panels by definition. These, so-called Conventional Profiling panels, are trained with respect to a fixed set of attributes.

Because the distinction between different types of sensory and consumer panels is not always clear, and because the data that result from all profiling-type panels are not very different, both terms *sensory* and *consumer* research appear in this book. Another reason for this is that the Multivariate Analyses applied can be used for both Sensory and Consumer data. As a result, when the term *sensory research* is used it can be read to mean *sensory and consumer research*.

1.4 Sensory Profiling

A large number of sensory studies are of the sensory *profiling* type. There are two different kinds of profiling studies: *Conventional profiling* studies and *Free Choice Profiling* studies (Williams and Langron 1984, Williams and Arnold 1985). The data from either profiling method are usually derived from the position of marks along a line scale or scores on a rating scale or from a category scale. The assessor marks his/her perceived intensity of an attribute on a line scale or indicates the appropriate category of a category-scale. Figure 4 gives an example of four line-scales for four attributes.

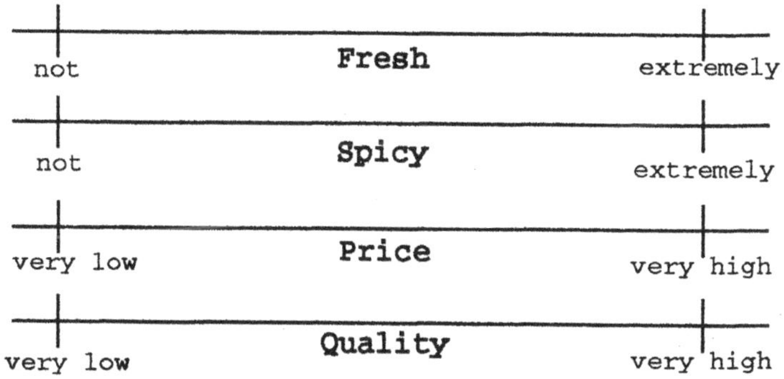

Figure 4 Example of four line-scales, for the attributes fresh, spicy, price and quality.

Figure 5 shows two examples of another type of scale, the category scale. These scales have a limited number of categories of which the assessor can choose one. A comparison of the results of using line-scales and category-scales can be found in Chapter 6 (van der Burg and Dijksterhuis 1993). A disadvantage of that study is that the line-scale data were converted into a low number of categories *a posteriori* (see also Chapter 5, van der Burg and Dijksterhuis 1989). In this way the effect of a different response behaviour of the assessor, resulting from the presentation of a different kind of response scale, is excluded from the study. It would be interesting to study this particular aspect of the differences in use of response-scales.

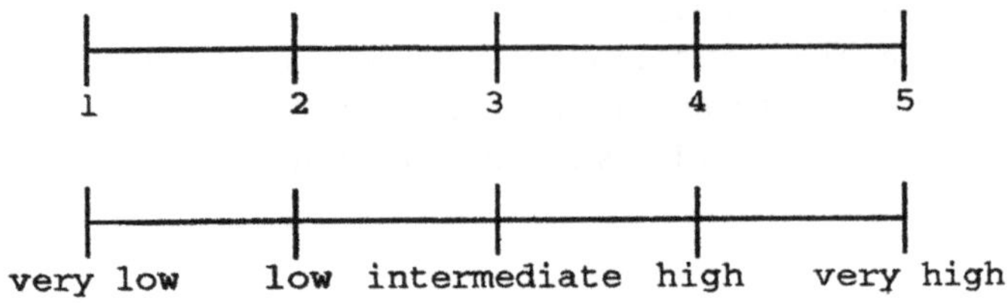

Figure 5 Two different category scales, a 5-point numerical and a 5 category adjective scale.

The type of response scale used is intimately connected to the problem of the measurement level and the admissible scale transformations of the data. This point is returned to in §1.6 and in Part II.

There are other types of response-scales too. King (1986) reports the use of an audio method in which the assessors give their scores by adjusting a tone to a certain pitch. Non-graphical response scales, as King's pitch-scale, deserve to be studied too. A disadvantage is that special devices are needed, and graphical scales are much easier to employ.

1.4.1 Conventional Profiling

In conventional profiling, a fixed vocabulary of descriptive terms is used by the sensory panel to judge the products. A sensory panel is often trained in the use of these terms. In the case of e.g. QDA (Quantitative Descriptive Analysis, see Stone & Sidel 1985), the panel starts with the generation of a lot of terms that are thought useful to describe the products under consideration. The whole procedure of attribute generation and training may take months. It is assumed that all assessors are able to use the attributes in the same way, so individual differences in use of the attributes are minimised due to the training. When one assumes no individual differences or ascribes them to noise or random error, individual judgements can be averaged and e.g. Principal Component Analysis can be applied to the average scores.

The data from conventional profiling experiments can be seen as a 3-mode data structure built from N products, M attributes and K assessors (see Figure 6).

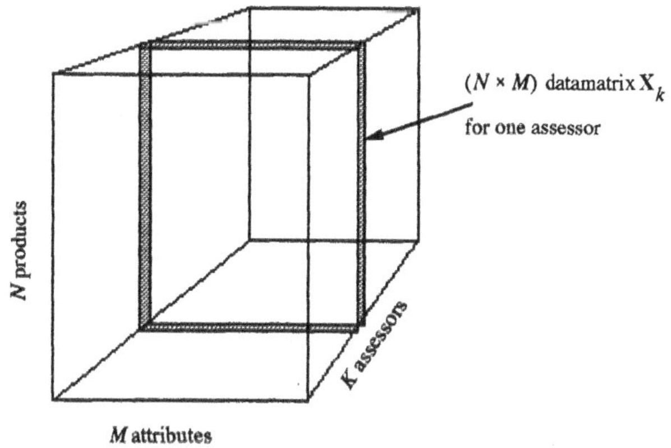

Figure 6 3-mode data structure representing Conventional Profiling data: N products are judged by K assessors using M attributes.

1.4.2 Free Choice Profiling

In Free Choice Profiling (FCP, Williams and Langron 1984, Arnold and Williams 1985), the assessors are free to come up with their own attributes, which they use for judging the products. So there is no *a priori* agreement on attributes between the assessors. As a result, it is impossible to average the individual data directly, because it makes no sense to add different attributes. The data from Free Choice Profiling experiments *must* be analysed by individual difference models which come up with some kind of average after transformation of the data. Unlike Conventional Profiling data, Free Choice Profiling data cannot be arranged in a kind of 3-mode data structure because each assessor $k=1,\ldots,K$ may have a different number of attributes (M_k). More importantly, the jth attributes of the assessors are not necessarily the same. Figure 7 illustrates the structure of an FCP data set.

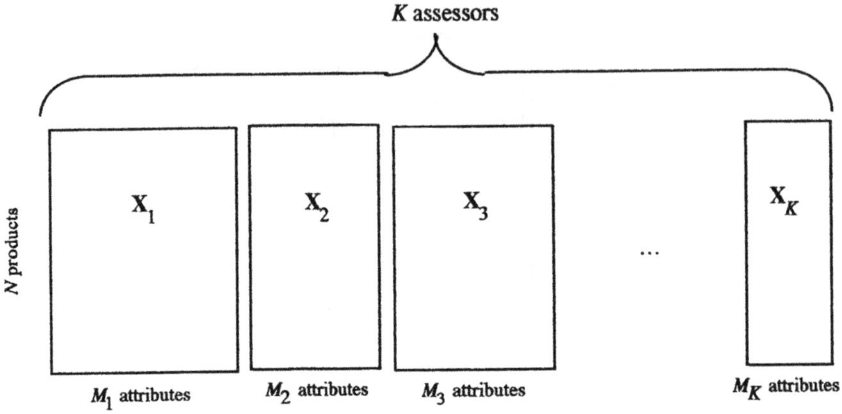

Figure 7 Data structure representing Free Choice Profiling data: N products are judged by K assessors each using M_k attributes.

Figure 7 shows that the individual data matrices $\mathbf{X}_k$ cannot be arranged such that the attributes match because each assessor's individual data matrix contains different attributes.

1.5 Individual Differences

Differences between the data of the assessors in a sensory panel are a concern in most sensory studies. Because in sensory research the chemical senses (smell and taste) play an important role, there are rather large individual differences between the judges. These differences may be larger than with the visual, auditory and other senses. The lack of consensus is for a large amount due to two effects, one physiological, and one psychological:

- large individual differences in the internal milieu of the chemical senses, i.e. the nose and mouth;
- there is no clear standard vocabulary concerning the sensations of taste and smell.

The first effect results in different perceived intensities of stimuli and different time courses of the perceptions. The differences in time course are found clearly in TI-studies (see also §1.8 and Part IV).

The second effect results in problems with the interpretation of the behavioural responses elicited from the assessors. The four basic tastes, sweet, sour, salty and bitter are clear, but flavours involve the sense of smell and there are no basic smells known. Everyone may use another term to describe the same sensation. This is the main reason that sensory panels are trained when exact and consistent sensory *analytical* data are needed.

Under the assumption of only a physiological effect, proper standardisation of scores should correct for much of the individual differences. In that case, individual scores could be averaged and analysed subsequently by e.g. PCA. When the psychological effect plays a role too, and it most often does, standardisation is not enough, and special methods that correct for the so-called interpretation-effect are needed.

When averages are computed over individuals, both the physiological and the psychological effect can be interpreted to give rise to random error only. But, when more elaborate data analysis is employed, as will be illustrated in this book, some of this error appears not to be random and may contain interesting information.

1.5.1 Subjects, Objects and Variables: Three-Modes and Three-Ways

A typical sensory profiling experiment consists of presenting a group of people, the *panel*, with a number of products and asking them to judge the products on a set of attributes. In more formal terms: *subjects* are presented with *objects* which they judge using a set of *variables*. The data resulting from such an experiment can be characterised as consisting of three ways, corresponding to the three *modes*: objects, subjects and variables (see also §1.4.1 and Figure 6). The data can be classified as three-way, three-mode data (Carroll and Arabie 1983). When K assessors judge N products on M attributes, the corresponding data can be presented as a three dimensional table (see Figure 6). An element x from such a three-way data matrix X can be identified by three subscripts:

$$x_{ijk} \in \mathrm{X},\ i=1,\ldots,N;\ j=1,\ldots,M;\ k=1,\ldots,K \qquad (1)$$

Such data are typically multivariate, at least it will be assumed they are (see Heiser 1992). For the multivariate analysis of this kind of data special three way techniques exist (see e.g. Law *et al.* 1984, Coppi and Bolasco, 1989).

1.5.2 *Averaging and Individual Differences*

A common way of analysing sensory data is by using averages. The first step in these analyses is the averaging of the individual data matrices. An individual data matrix is one slice X_k of order (N×M) from the three-dimensional structure in Figure 6. The average data matrix looks just like this slice of Figure 6, with the difference that it contains averaged scores, $i=1,...,N$; $j=1,...,M$, instead of individual data x_{ijk}:

$$x_{ij\cdot}=K^{-1}\sum_{k=1}^{K}x_{ijk} \tag{2}$$

The average data matrix $\bar{\mathbf{X}}$ can be analysed by means of Factor Analysis or Principal Component Analysis. The averaging of the raw data naturally results in loss of individual differences.

As an alternative to averaging it is possible to perform a PCA on all variables of the concatenated sets which amounts to an analysis of an $(N \times MK)$ data matrix. Such an analysis results in MK component loadings which can be inspected. In a plot the loadings from the same assessor can be marked for easy identification of which variable goes with what assessor. The disadvantage of this strategy is that the individual assessors may not be represented fairly. Weighting variables per assessor may help but eventually other methods will be more appropriate. To solve problems like these an individual difference model can be useful.

Three-way models offer a solution because they respect the third mode, here the different assessors, in the data. However, these models assume equality of variables over subjects. This assumption may be justified for data which contains clear and unambiguous variables but probably not for most sensory data.

1.5.3 *Sets in* K-*sets Analyses*

The assessors in a sensory panel are the measuring devices with which the data are collected. The human being acting like a measurement device can measure e.g. the shape, the colour, the apparent length, the taste, the smoky odour, and lots of other characteristics of objects. Each individual device (assessor) produces and uses these variables in its own idiosyncratic way. It is as if all devices were differently, and obscurely, calibrated, and it is unknown what it is they measure. This confusion is the reason that the attributes used by one assessor belong together and are distinct from the variables from another assessor. In terms of this study, they constitute a *set* of attributes. The application of GPA and GCA in this book is such that an assessor is represented by a *set* in the data. The individual assessor's set is transformed by Generalised Procrustes Analysis (or Generalised Canonical Analysis) to maximise the agreement between the assessors.

Because each set consists of its own attributes and does not necessarily contain the same attributes as the other sets, the data cannot be represented in a three-way table anymore (see §1.4.2 and Figure 7) since the third-way does not match. Data with variables grouped into sets like this is called *more sets* data, or *K-sets* data.

1.6 Measurement Levels

Line-scales are perhaps the most common measuring instrument in sensory profiling. The scores obtained with such scales are numerical and may range from 0 to 100, but the range is unimportant. It is usually assumed that the scores are interval or ratio-type and can safely be used in linear Multivariate Analysis models. When they range from 0 to 10, this assumption may be violated, and the violation may be worse, the less distinct scores there are. Category-scales are less often used in sensory science, perhaps because of the lack of appropriate statistical models, though Multidimensional Scaling methods (see e.g. Shepard *et al.* 1972, Young and Hamer 1987) can give interesting results (Schiffman *et al.* 1981, MacFie and Thomson 1984) despite the fact that the Gifi (1990) system of non-linear MVA has been available in a major statistical software package for some years now (SPSS 1990).

The second theme of this book concerns the problem of measurement levels of sensory data. The question is whether ordinal analyses of a low number of scores give better results than the usual linear analyses, or perhaps it can be shown that it does not make much difference whether the scores are analysed linearly (with an assumed numerical measurement level) or non-linearly (with an ordinal measurement level or a nominal measurement level assumed). See Chapter 6 (van der Burg and Dijksterhuis 1993b) for such a comparison.

When one realises what an assessor does in a sensory experiment, it need not be a surprise that non-linearities occur:

tasting / smelling → perceiving taste / smell → perceiving distances on line - scale or between categories → judging → scoring

In this simple model, the assessor switches from sensory tasks (tasting/smelling and perceiving distances) to a judging task (matching distance to the intensity of the taste/smell) to a motor task (marking a score). Non-linearities are indeed encountered in sensory data and can be modelled using non-linear data analyses methods.

1.6.1 Non-Linearities in the Data

As is since long known from psychophysics, the relation between a physical stimulus and the perceived intensity of this stimulus is not linear but rather logarithmic.

Weber's law (or the *Weber-Fechner* law) is written

$$\psi = k \log \Phi \tag{3}$$

With Φ the physical stimulus intensity, k a constant and Ψ the perceived stimulus intensity (see Figure 8).

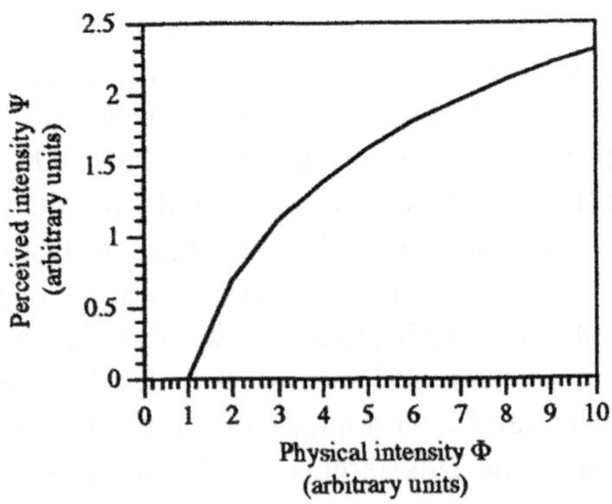

Figure 8 Weber-Fechner logarithmic law.

Stevens (e.g. 1962) proposed a Psychophysical Law in which the logarithm is replaced by a power function. This so-called Power Law is

$$\psi = k\Phi^n \tag{4}$$

where n is the exponent of the power function. Different modalities give rise to a psychophysical function with a different exponent. When the exponent $n=1$, a linear function results. Figure 9 shows some power functions with exponents $n \in \{0.1, 0.5, 1, 2, 5\}$.

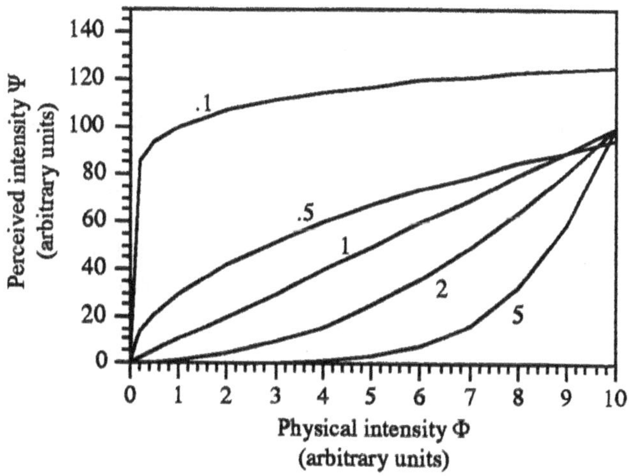

Figure 9 Stevens' power law with different exponents n and constants k (k was chosen to make the function fit the frame and to show its most non-linear part, for illustration's sake only).

Over the years, a lot of exponents of power functions have been collected in a large number of psychophysical experiments. Table 1 lists some exponents for a number of smell and taste stimuli found in the literature (see Dember and Warm 1979, p. 93, Table 4.1).

Table 1 Exponents of power functions relating physical intensity to perceived intensity of some smell and taste stimuli (after Stevens 1960).

modality	stimulus	exponent *n*
smell	coffee odour	0.55
	heptane	0.6
taste	saccharine	0.8
	sucrose	1.3
	salt	1.3

Power functions with the exponents in Table 1 are drawn in Figure 10. This figure is drawn to illustrate the apparent non-linear relationship between the physical and perceived intensity of the stimulus. When another range of the physical intensity is selected, the functions may be reasonably well approximated by a linear function. In practice however, stimuli near the lower threshold, i.e. with low physical intensities, as well as stimuli with high physical intensities, will be encountered so it is unknown whether linear approximation will be satisfactory. Especially with sensory profiling of real food stuffs, contrary to controlled model solutions, the physical intensity of most attributes is unknown. Often even the precise chemical or physical cause of certain attributes will be unknown.

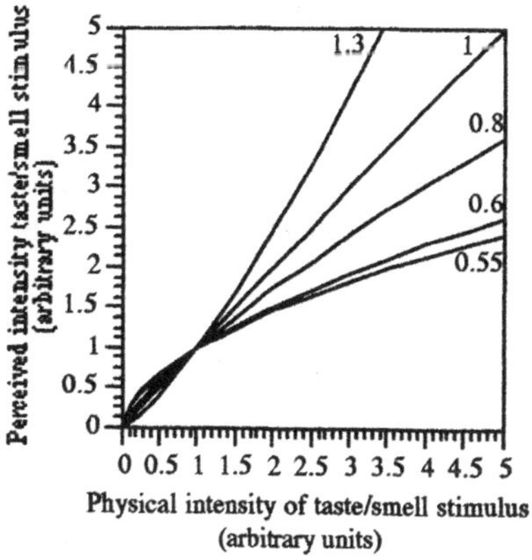

Figure 10 Power functions with exponents from some taste and smell stimuli.

When instead of perceived intensity, preference (acceptance or liking) is measured and its relation with physical intensity of the stimulus is plotted, non-linear relationships are very likely to occur (see Figure 11).

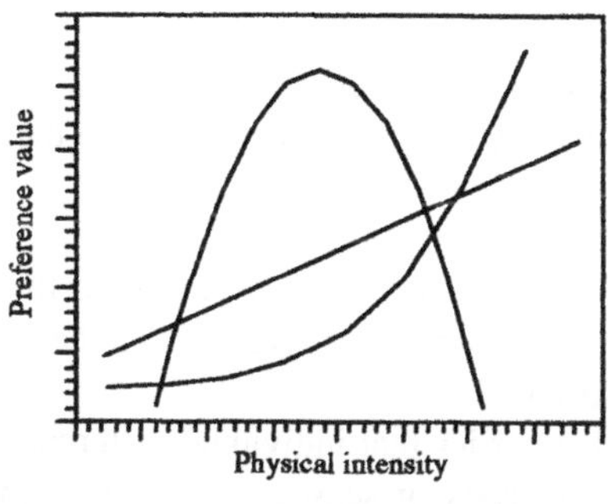

Figure 11 Theoretically possible relationships between the physical intensity of the stimulus and the perceived preference value of the stimulus.

In sensory studies, the physical stimulus can be e.g. different levels of sweetener, and the behavioural response a preference-score. In this case the inverted U-shape in Figure 11 may be encountered. Other stimuli will have differently shaped preference functions. Most will be inverted U-shape-like ("sweetness", "saltiness"); very few will be linear. It is clear that *appreciation*, ("preference", "acceptance", "liking", etc.) data is probably not modelled using linear approximations.

Other attributes will have non-linear relationships too. As an illustration Figure 12 presents three paired-scatter plots of three attributes from one assessor from a data set with 120 products.

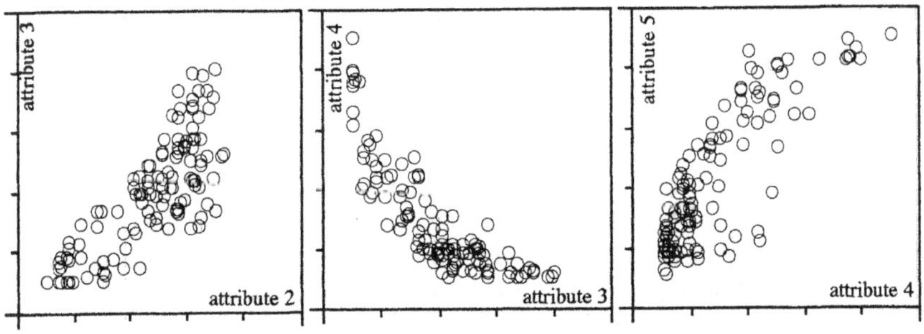

Figure 12 Relationships between three different attributes.

The relationships in Figure 12 are from the data from one assessor. One could comment that it may be preferred to use average data to be approximated by linear models but:

1. when all assessors show such a clearly non-linear relation between attribute 3 and attribute 4 in Figure 12, the average assessor will probably do so too;
2. it was concluded earlier that individual difference models are a useful device for the analysis of sensory data, so no averaging takes place (and with FCP-data averaging is impossible).

It may well be that in practice linear relationships are the exception rather than the rule. It is non-linearities like those illustrated in Figure 12 that play part in all kinds of profiling data, be it conventional or free choice profiling.

Another subject where non-linear relationships occur is in the study of Sensory-Instrumental relations, the topic of §1.7.

1.6.2 Non-linear Treatment of Categorical Data

Another source of non-linearities is found in the way the assessors use categories. It is commonly assumed, but often not justified, that the assessors use categories as numerical (ratio) data. This would mean that, say, a sweetness judgement of 4 means that the stimulus was perceived twice as sweet as one with the judgement 2. An ordinal relationship between the categories of a category-scale would perhaps be closer to the truth. A score of 4 is more than a score of 3, which is more than a score of 2, etc. It is unspecified exactly how much more it is. Another possibility would be that 4 could well be meant to be less than 2, and 3 in between. In the sweetness example this would probably not apply, but with preference scores this may not be uncommon.

When numerical scores from line-scale variables are converted into categories, or directly collected as categories from a category-scale, non-linear (nominal or ordinal) analysis of the data may be useful. In Part II this topic will be studied in more detail.

1.6.3 Non-linearities in MVA

In Figure 13 an example of non-linear transformations of a number of categories, say from a 10-point category scale, is given. Note that the figure is made purely for illustrative means. The transformed data in the panels in Figure 13 are fictitious. The figure contains two variables, x and y, which clearly have a non-linear relation (leftmost panel). When the categories of x and y are transformed ordinally the relation becomes somewhat more linear (middle panel). The categories are indicated along the x-axis. It shows in the unequal spacing between the categories that they are transformed non-linearly — they are not spaced equally. In the rightmost panel a nominal transformation is illustrated. In addition to the spacing between the categories, the order of the category-numbers along the x-axis has changed. The same transformations are applied to y too (no category-numbers were drawn for y in Figure 13). It is also possible that x and y receive different transformations. In Figure 13 only two variables are shown, but in practice transformations are applied to all variables in the analysis.

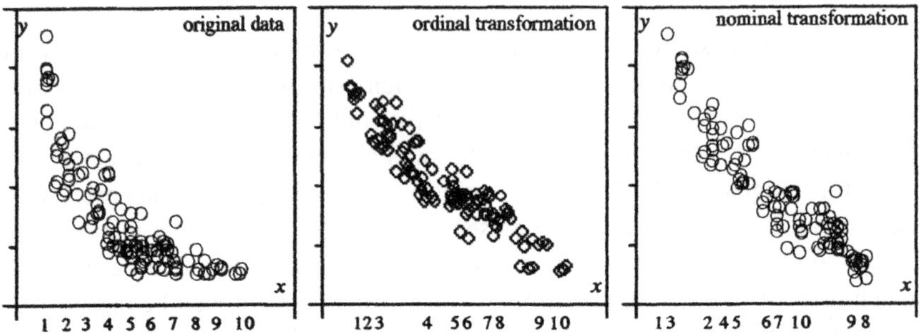

Figure 13 Illustration of non-linear transformations of two variables in a non-linear MVA with optimal scaling.

With the usual linear analyses, a lincar relationship is imposed onto the data. Imagine this for the data in the leftmost panel of Figure 13. A linear relation may be inappropriate, though it is recognised that it may often provide reasonable approximations (see Heiser and Meulman, 1993, p. 1).

The process illustrated in Figure 13 is called *optimal scaling* (Young 1981). For two variables (as in Figure 13), the process effectively linearises the regression of x and y. In the Gifi (1990) system of non-linear Multivariate Analysis an optimal scaling step and a linear MVA step are alternatingly performed until a certain criterion is satisfied. This procedure is known as Alternating Least Squares, hence the suffix *ALS* of the Gifi-methods (Homals, Princals, Canals, Overals, etc.).

1.6.4 Individual Differences and Measurement Levels

The way numerical scores are used can differ between the assessors in a sensory panel. This is why the application of methods that combine an individual difference approach with a non-linear (i.e. nominal or ordinal) analysis is interesting. In Chapter 5 (van der Burg and Dijksterhuis 1989) an analysis is presented which shows that different individuals received a different quantification of their category-scores. It reflects a different use of numerical scores. In that study the low number of categories was constructed *a posteriori* from line-scale scores which is a methodological disadvantage. It would have been better if two different experiments had been carried out, one with line-scales and one with category-scales.

1.7 Sensory-Instrumental Relations

The third theme in this book is the study of Sensory-Instrumental relations. The idea behind the study of Sensory-Instrumental relations is that sensory perceptions have chemical/physical counterparts in the substance under investigation. A simple example is the amount of caffeine in a drink, which determines the perceived bitterness. In real life Sensory-Instrumental research is much more complicated, and can involve complicated multivariate data from different sources (see e.g. the "Understanding Flavour Quality" symposium, 1992). Consequently, Multivariate Data Analysis finds an interesting field of application here.

1.7.1 Sensory-Instrumental Data

In Sensory-Instrumental studies, one data set ($\mathbf{X}_1$) contains the sensory judgements on a number of products (say *N*). Another data set ($\mathbf{X}_2$) contains a number of instrumental measures on the same *N* products. These can be results of chemical analyses, physical properties and of other measurements on the products. An illustration of the two data sets involved in Sensory-Instrumental data analysis is given in Figure 14.

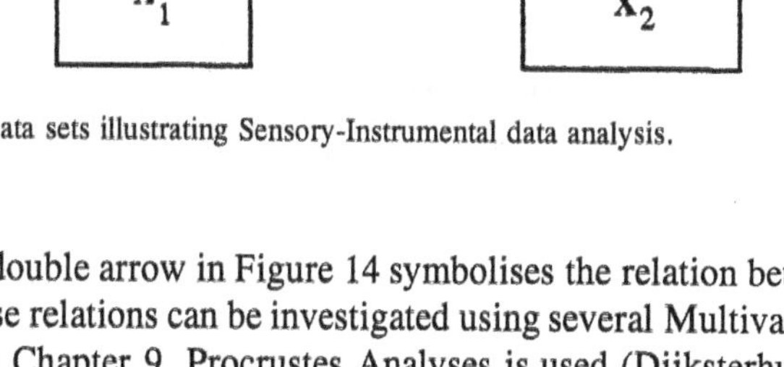

Figure 14 Two data sets illustrating Sensory-Instrumental data analysis.

The double arrow in Figure 14 symbolises the relation between the two data sets. These relations can be investigated using several Multivariate Analysis techniques. In Chapter 9, Procrustes Analyses is used (Dijksterhuis 1993b). In Chapter 7 and 8 Redundancy Analysis and Canonical Correlation Analysis are used to this end (van der Burg and Dijksterhuis 1992, 1993a). Note that each of the sets can be the result of prior analyses. The sensory set, $\mathbf{X}_1$, may very well be the group average from a Procrustes Analysis of the data from a sensory panel.

The multivariate methods to study the relations between the two data sets can differ in three respects:

1. symmetry;
2. measurement level;
3. criterion.

These three topics are the subjects of the following three sections.

1.7.2 *Symmetric and Asymmetric Analysis of Two Data Sets*

The methods that relate two data sets can be classified into two types: *asymmetric* methods and *symmetric* methods. The symmetry concerns the way the two data sets are treated by the method. *Asymmetric* methods try to predict one set from the other, and so treat both sets differently. Partial Least Squares regression, Principal Component Regression, Redundancy Analysis and Multiple Regression are among these methods (see Figure 15).

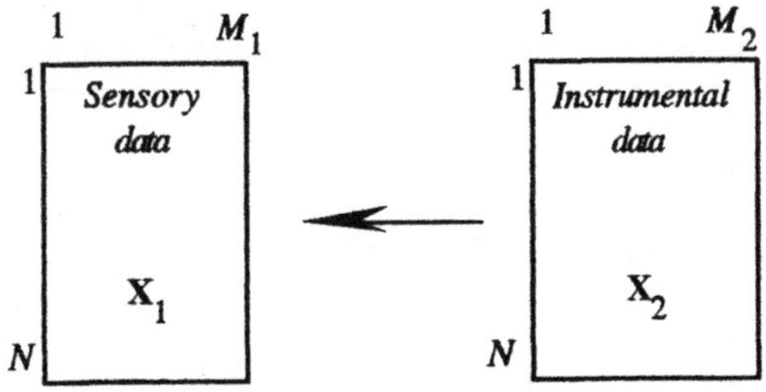

Asymmetric methods: **PLS**
Principal Component Regression
Redundancy Analysis
(Multiple) Regression
(M)ANOVA

Figure 15 Two data sets illustrating asymmetric data analysis models.

When both the set X_1 and X_2 contain one variable, $M_1 = M_2 = 1$, ordinary regression results. When X_2 contains a design-set, i.e. binary (dummy) variables coding an experimental design, a MANOVA method results. An example would be the crossing of three levels of sweetener, four different drinks with three temperatures, resulting in 10 dummy variables in X_2. When the $3 \times 4 \times 3 = 36$ stimuli are judged on perceived sweetness, only ($M_1 = 1$) an ANOVA can be performed. When other attributes are also used ($M_1 > 1$), a MANOVA must be applied.

Symmetric methods treat both sets identically. Swapping the two sets makes no difference. Neither of the sets is the object of prediction; only the relations between the sets are studied. Examples of these methods are Canonical Correlation Analysis and Procrustes Analysis (see Figure 16).

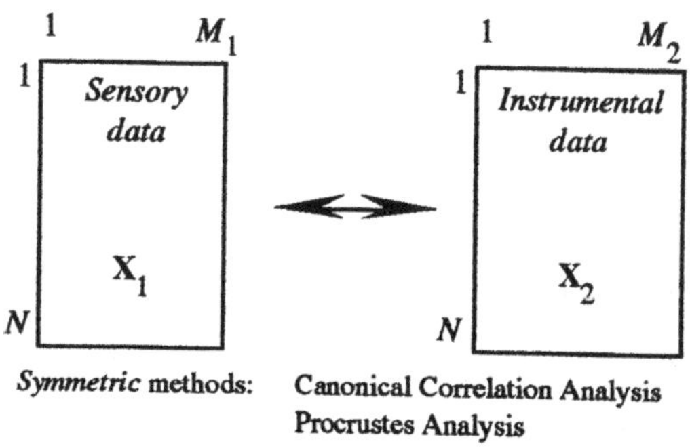

Figure 16 Two data sets illustrating symmetric data analysis models.

1.7.3 Non-linearities in Sensory-Instrumental Analysis

The examples of non-linear relations between variables in §1.6.1 are valid for Sensory-Instrumental relations too. Especially when studying appreciative sensory judgements one has to beware of non-linear relations (see Figure 11, Noble 1975). On the other hand, there is a risk of overfitting when granting the matching method too much freedom in matching the two data sets. The balance between imposing linear restrictions, with the risk of missing interesting relations, and imposing hardly any restrictions, with the risk of fitting noise, may be hard to find.

1.7.4 Criterion of Relation Between the Two Data Sets

The arrows between the data sets X_1 and X_2 in Figure 14, 15 and 16 symbolise the "relation" between the two data sets. This relation can be defined in different ways, and gives the difference between the resulting methods. The methods of ordinary linear regression, Multiple regression and Canonical Analysis, as well as ANOVA and MANOVA, use essentially the same criterion. They could be interpreted as belonging to the same family of two-data sets methods. Redundancy Analysis, Principal Component Regression, Partial Least Squares regression and Procrustes Analysis use different criteria. It is beyond the scope of this introduction to define these criteria. The papers in part III give more information on the criteria of the methods employed there, or give references to relevant literature. The methods used in these papers are presented in Table 2, where their main differences in symmetry, measurement level and criterion can be found.

Table 2 Differences in the three methods used in the chapters in Part III with respect to their symmetry, measurement level and criterion.

Chapter	method of analysis	symmetry	measurement level	criterion[1]
7/8	Redundancy	asymmetric	ordinal/numerical	maximal covariances
8	Canonical Correlation	symmetric	ordinal	maximal correlation
9	Procrustes	symmetric	numerical	minimal variances

2 This is a very crude description of the criterion; it is used to illustrate the differences between methods. A more elaborate exposition of criteria can be found in the references in the corresponding chapters.

1.8 Time-Intensity Data Analysis

The fourth and final theme in this book is Time-Intensity studies. In Time-Intensity studies, a taste stimulus is given to an assessor and her/his task is to try to track the perceived intensity of an attribute, e.g. bitterness, over time. For an overview of TI-research see e.g. Lee and Pangborn (1986) or Punter *et al.* (1989). A special computer program can be used to this end (e.g. Yoshida 1986, OP&P 1991, see also Dijksterhuis and Roos 1990). The assessor is shown a slider on a screen and she/he can move the slider using a computer mouse. A so-called *Time-Intensity Curve*, which is a graphical representation of the recorded intensity against the recording time, results (see Figure 17). In general, assessors are able to track the change of the taste rather accurately, but there are large differences between subjects (see e.g. Overbosch *et al.* 1986, Flipsen 1992, van den Broek 1993).

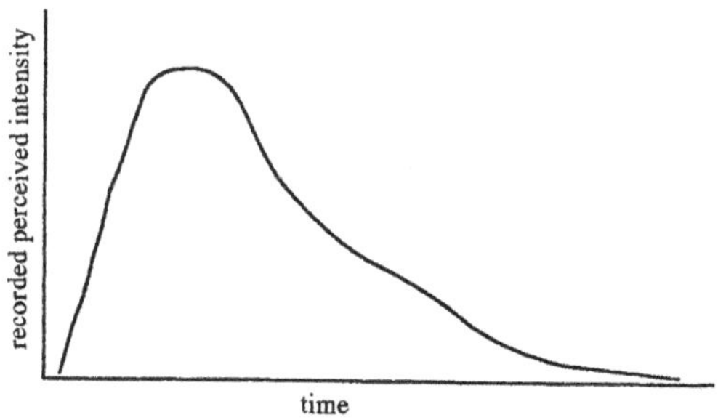

Figure 17 Typical Time-Intensity curve.

The original analysis of Time-Intensity (TI-) curves focuses on TI-curve parameters such as e.g. maximum intensity; time of maximum intensity, area under the curve, steepness of the flanks of the curve, etc. The parameters are often inferred from an averaged TI-curve. The averaging of TI-curves is one of the things criticised by some researchers and alternative ways of aggregating the TI curves are proposed (Overbosch *et al.* 1986, Liu and MacFie 1990, MacFie and Liu 1992, van Buuren 1992). Chapter 10 and 11 present analyses in which a weighted average is computed and individual curves are better represented in this weighted average curve (Dijksterhuis 1993a, Dijksterhuis *et al.* 1994). In

Chapter 12 a suggestion for an alternative kind of analysis of TI curves is given, based on the *shape* of the TI-curves (Dijksterhuis and van den Broek, 1994).

The object of TI-studies is the change of the perceived intensity of a flavour attribute, often bitterness, but other attributes are studied as well (e.g. sweetness, fruitiness). It would take too long to include an introduction to TI research here, but the interested reader is referred to (Nielson 1957, Larson-Powers and Pangborn 1978, Dijksterhuis and Roos 1990, Punter *et al.* 1989). Lee and Pangborn (1986) provide a list of references to applications of TI-techniques. The chapters in this part contain only a brief introduction to TI-studies.

The data that result from TI-experiments have some particular properties:

1. TI-studies often generate a large amount of data
2. there are clear intra-individual consistencies
3. visual inspection shows clear differences between stimuli and between different attributes
4. there are large inter-individual differences
5. TI-data always have a distinctive shape: the typical TI-curve is unimodal, skew, and levels off gradually, etc. (see Figure 17).

It is not uncommon that intensity scores are collected each second, during two minutes, so one TI-curve consists of 120 intensity scores. In a typical TI-experiment, some 12 assessors may judge eight different stimuli, perhaps even replicated three times, resulting in 288 TI-curves (and 34,560 data-points). The first of the five above mentioned properties is amply illustrated by this example. It is obvious that a data-reduction method is mandatory. The second and third properties are indications that there is something in the data worth looking for, although the fourth property may suggest otherwise on first inspection of the data.

The fifth property is the starting point for the study of the properties of (the shape of) the TI-curve in relation to the stimuli and what is known of the perception apparatus and processes. This specific TI-curve shape (see Figure 17) recently called for some alternative methods of TI-data analysis, mentioned in the Concluding Remarks of Part IV. The properties listed above give a data-analyst the feeling that there is probably more than meets the eye in those TI-curves. Part IV of this book expands upon this.

1.9 Data Analysis, Confirmation and Exploration[3]

In this book the use of Multivariate Analysis is of an exploratory nature. There are several reasons for this. The type of research, type of data, and above all the type of questions asked, are such that model-based statistical hypothesis testing is probably not the way to go. Such statistical decision making presupposes *a priori* stating of hypotheses, random sampling from some well-specified population, the construction of an experimental design for the experiment and, after collecting the data, the confirmation or refutation of the hypothesis. The testing is done with an *a priori* fixed level of α, the probability of making a type I error, i.e. the probability of rejecting the hypothesis when it is actually true.[4] When the hypothesis is rejected, a new hypothesis is deduced from a possibly adjusted theory, and a new experiment done. This rather strict way of statistical hypothesis testing may not be practical for making progress in sensory and consumer research. Especially in applied contexts other methods may be preferred.

1.9.1 Confirmatory and Exploratory Mode of Analysis

Some well known methods of data-analysis are ANOVA and MANOVA (see e.g. Hand and Taylor 1987). These are often presented as confirmatory methods in the sense that *a priori* hypotheses (effects) are tested by means of designed experiments and subsequent analyses. However, in particular MANOVA methods can grow very complex, and as a result are liable to become exploratory rather than confirmatory methods, which is not necessarily a bad thing. Since most MVA methods are not purely exploratory or confirmatory, it is preferable to think in terms of an *exploratory mode* of analysis versus a *confirmatory mode* of analysis. In any case, the data are collected using an experimental design for minimising certain unwanted effects (Cochran and Cox 1957)[5] and stressing the effect under study. In exploratory mode, the data are of primary concern, and models are merely suggestions, or come in later. If we "see" an effect, it is very useful to be able to state the probability that it could have occurred by chance, and this is exactly what significance tests do for us. Standard parametric significance tests still assume random sampling, but non-parametric permutation tests can be performed

[3] Parts of this section are taken from Dijksterhuis and Heiser (1995).

[4] Of course the power 1-β is important too, but the subject is too vast and complicated to be covered here.

[5] For sensory applications of some experimental designs see MacFie *et al.*(1989) and Schlich (1993b).

without that assumption; they merely prescribe the random sampling from certain permutations of the data.

A disadvantage of model-based statistical procedures is that a lot of assumptions must be fulfilled to be able to test the hypotheses or the models. In practice, the assumptions are seldom checked, sometimes ignored, and as a result the validity of the obtained results may be low. An example is the multiple testing of hypotheses. When testing more than one effect from the same data, the level of α should be adapted, which is seldom done in practice. In confirmatory mode, models are of primary concern, and the data are secondary, except for the aspects that distinguish the models. Indeed, model testing is the prime instrument for confirmation. Yet confirmation is a process that would soon depend on prejudice alone if we were not allowed to listen to unexpected signals from the data. These signals can be the seeds for new hypotheses to be confirmed in fresh observations.

Exploratory -multivariate- data analysis is not less "strict" as may be thought by some. In many cases one can say that MVA is just transforming the data in one way or another, and inspecting it from another angle.[6] The transformations are not arbitrary, they are strict mathematically defined transformations, sometimes approximated by algorithms, but strict they remain. The exploratory, and potentially subjective, part is in the conclusions that the researcher draws from the transformed data.

1.9.2 Why is Formal Statistical Inference Rare in Descriptive Sensometrics?

There are some reasons that formal statistical inference is not very common in (Multivariate) Sensometrics. Below the situation concerning Principal Component Analysis is illustrated, but the results hold for other MVA methods too.

A statistical test to choose the dimensionality in which to present the results of a PCA, would be a welcome tool. Higher dimensions than the selected dimensionality would be attributed to noise, and the dimensions 1 through p would be assumed to model signal. But here we encounter, not surprisingly in a rapidly developing discipline, an embarrassment of riches. In the literature, several methods are proposed to infer this dimensionality p. Some of these methods are:

---

[6] Sometimes the data are literally inspected from another angle, e.g. after rotations or projections.

- choose all eigenvalues greater than unity
- Bartlett test
- split-sample procedure
- scree-test.

The merits of these methods are lucidly commented on by Cliff (1987, Chapter 13; see also Zwick and Velicer 1986). It is concluded that the often used eigenvalues-greater-that-unity-rule should not be taken too seriously. Sometimes it underestimates the number of components, especially with *post hoc* rotations. In other cases it overestimates the number of components. Bartlett's test may accept too few or too many components depending on the size of the data matrix. The split-sample approach is an interesting, though seldom used, method. It involves extra computations, which, by the way, is not an argument anymore not to employ a technique since personal computers have the power of the mini's of ten years ago. These split-sample methods, however, do presuppose a large data matrix, in particular a large number of objects *N*. In practice *N* may be too small to employ this method safely. Looking for an elbow in a scree graph is a very easy and useful method. But this scree-test is not formally a *test*, and is not devoid of misinterpretation.

Another approach is to analytically study the distributions of eigenvalues and their sensitivity to disturbances (see e.g. Krzanowski 1984). These methods suffer from several drawbacks (see e.g. Jolliffe 1986, p. 39–41):

- the mathematics is complicated
- the results are often asymptotic
- the original data are often assumed to be multivariate normally distributed.

The disadvantage of the first point is that it will take time and a lot of work to put the results into practical use, e.g. by means of software implementations, and the methods may not find common application as a result of their complexity. The second point is perhaps more serious. Often the data matrices involved are of rather small size, and it is not known to what amount the results will apply under "non-asymptotic", but highly realistic conditions. The third point is also important. Real data may often not be multivariate normally distributed, so this assumption is often violated. In addition, with nonlinear models, the distribution of the residuals may be normal without leading to a normal distribution of the data, or the other way round. Since a good deal of MVA methods are nonlinear, this fact complicates the inference process enormously. In view of these reasons, it need not surprise us that formal, model-based statistical inference is often omitted in Multivariate Sensometrics.

1.9.3 Implicit Experimental Design

Some may argue that exploratory multivariate data analysis suffers from a lack of experimental design. This criticism is heard in particular with respect to sensory profiling studies, where e.g. GPA is used for the analysis of the data. However, Experimental Design *is* present in such studies, albeit perhaps not explicitly. The design is implicit in the analysis of the data. The following "design" choices are made when the data are analysed, e.g. by means of GPA, but probably the argument holds for any multivariate data-analysis method:

- choice of observation-units (products), judges and attributes
- choice of representation
- choice of test

The choice of observation-units is the first design-choice made. In a sensory profiling experiment it involves selecting the set of products, whom to present it to and which attributes will be used. In data-analytical terms, these are the *modes* of the future data matrix. The choice of representation determines which *mode* will be allotted to which *way* of the data matrix (for *modes* and *ways* of a data matrix see Carroll and Arabie 1980). In sensory profiling experiments the assessors will be represented in the *sets* in the data analysis, the products in the rows, and the attributes in the columns of the datamatrix.[7] The third choice is the choice of test. This is effectively the choice of the options within a broad class of analysis methods available for data with the chosen shape and the chosen representation. Though at first sight this choice will perhaps not be recognised as a design-choice, the choice of options reflects the *implicit* design just as the *explicit* sampling design follows from the statistical test or from the (M)ANOVA model employed. For example, GPA is used to find the agreement between the *sets* (often assessors), under translation, rotation and isotropic scaling of the configurations of row-objects (products) within sets. GCA finds agreement between sets focusing on the maximal correlation between linear combinations of the columns (attributes) of the sets.

---

[7] *K*-sets data differ from three-way data (see §1.4.2), but the distinction is ignored here to keep the reasoning clear. However, the argument holds for *K* sets data too.

1.9.4 Brief Ontogeny of Data Analysis Techniques

The three design-choices will probably not be made consciously by all researchers. When a data analysis technique is originally developed, its developer's intentions are to solve a particular data-analytical problem. The technique is tailored to solve that specific problem. The developer explicitly made the three design-choices to devise the technique. Subsequent researchers probably recognised similarities between the data-analytical problem of the developer and their own, and decided to use her/his technique. When they succeeded in solving their problem, they would report this in a scientific publication, talk about it to colleagues, present results at symposia, etc. Then the technique becomes known as a tool for the solution of a particular data-analytical problem. Later users of the technique will only use it because their data seems fit for this technique: it has become a paradigm. They will not explicitly make the three choices, and they don't always need to. Over time, the technique will become increasingly utilised as a black-box, where data is input, and the results are interpreted. Provided there is a clear manual[8] with the technique, this black-box use of techniques can be very useful. The manual should contain unmistakable instructions, preferably with reference to the three choices, about what problems the technique can be used for.

The above illustrates that the design, *explicit* for the initial developer of the technique, becomes more and more *implicit* for later "black-box-users" of the technique. It is preferred that the latter users know what the choices once were. When in doubt, researchers should consult an expert or decide not to use this technique.

1.9.5 Graphical Representations and the Eye

In the exploratory use of Multivariate Data Analysis, graphical representations are important. The author believes that a good researcher who knows her/his subject matter has an *eye*[9] very well equipped for recognising the signal from the noise, so to speak. This *eye* would have been capable of stating good research hypotheses and designing proper experiments for statistical hypothesis testing. The difference is that with the latter the *eye* is separating the signal from the noise before the experiment, and with exploratory MVA the *eye*

---

[8] The use of the word "manual" seems to imply that the technique is available as a computer program. Though this will often be the case, it is not necessary.

[9] Of course the brain connected to the eye is more important, but the italic printing of eye is meant to include the brain and "its" knowledge.

is doing it afterwards. By the way, the MVA techniques were also devised *before* the experiment, as ideally is the choice of variables to be analysed (so we cannot add or delete variables until the results seem to suit our purposes).

1.10 Structure of the Book

All chapters in this book, except the first and the last, were originally articles that have been published. The sections 1.5 to 1.8 introduced the four themes which are presented in the following four parts (see also Figure 18):

Part I Individual Differences (§1.5);
Part II Measurement Levels (§1.6);
Part III Sensory-Instrumental Relations (§1.7);
Part IV Time-Intensity Data Analysis (§1.8).

Each part is preceded by an Introduction and ends with Concluding Remarks. In the Introduction of a part the articles from which the chapters originated are referred to and some background information is provided. In addition, credit is given to colleagues, co-authors and others who helped with a particular paper. In the Concluding Remarks, the conclusions from the chapters in the part are summarised and, where possible, put in the larger framework of a particular problem in sensory research. Furthermore, possible future directions for the research under consideration are outlined.

Two of the four themes (I and II) are concerned with Sensory Profiling data, an important and abundant type of data in sensory research. Theme III-studies produce their own type of data of which the Sensory part can be from profiling studies, the Instrumental part stems from non-sensory, i.e. chemical or physical measurements. TI-data, theme IV, are yet another type of data. They are micro time-series of one or two minutes length. The different types of data will be illustrated in the corresponding chapters and their introductions.

The articles are arranged into four parts in this book as shown in Figure 18. There is some overlap between the themes, some of the chapters apply to more than one theme.

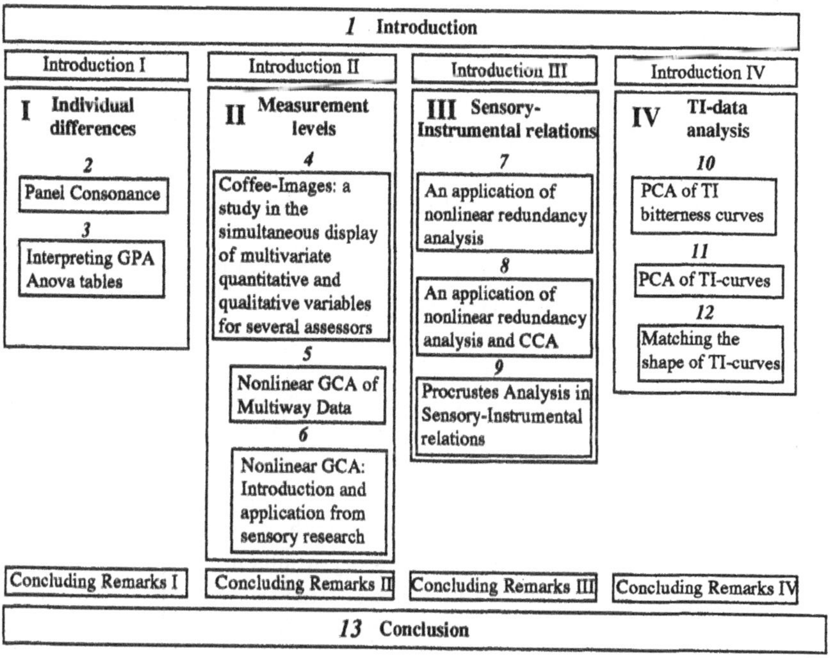

Figure 18 Structure of the book, the relation of the chapters (italic numbers) to the main themes in the parts (roman numbers I, II, III and IV).

Because each chapter was originally written as a more or less self-contained article, some matters will be introduced more than once. The reader is invited not to see this repetition as a problem because it can often be clarifying to re-read material in another context, in another style, by another author, etc. The advantage of the self-containedness of the chapters is that it enables reading individual chapters. Some chapters may be known to students of sensory science since they have been published in typical sensory journals as Food Quality and Preference and the Journal of Sensory Studies, while other articles, the ones with a somewhat more methodological content, have been scattered throughout several conference proceeding books and methodological journals.

1.10.1 Two Books or One?

This book can be viewed to consist either of two books or of one, depending on the main interest of the reader. Sensometricians will read it as one integrated book in which new multivariate methods are presented and shown to be fit for certain kinds of sensory data. For others there may be two books:

1. sensory problems, sensory data and some results obtained by MVA
2. the presentation of some recently developed MVA methods illustrated by the analysis of sensory data.

Readers with a mainly sensory interest may find some of the statistical parts too technical. The findings may be of use to them, but they perhaps need not know the details of the methods. For them Table 3 contains the sections they can read to find the results most relevant to sensory problems. They read *book 1* when they follow this lead. Readers with a mainly statistical and data-analytical interest may find some sensory results too elaborate. They can read book 2, following the lead in Table 3. They are invited to pay special attention to the references, which contain most of the original papers about the different MVA methods. All introductory parts as well as the Concluding Remarks are best read by both types of readers. They provide useful background information, a preview on the paper and summarise the main findings along with some indications for future research, both sensory and data-analytical.

Table 3 Two books, one with a sensory focus, one with a data analytical focus. Sections omitted from the table should be read by both kind of readers.

sections	*book 1*: sensory	*book 2*: data analysis
Chapter 2	√	
3.3-3.4		√
3.5-3.6	√	
4.2	√	
4.3		√
5.2-5.3		√
5.4	√	
6.2-6.3		√
6.4-6.5	√	
7.2-7.3		√
7.4-7.6	√	
8	√	
9.3		√
9.4-9.5	√	
10.2	√	
11.2	√	
11.3		√
11.4-11.5	√	
12.2		√
12.3	√	

PART I

INDIVIDUAL DIFFERENCES

INTRODUCTION TO PART I

Summary

In this part, the first theme of the book, Individual Differences, is studied by means of two Multivariate Analysis methods: (Q-mode) Principal Component Analysis and Generalised Procrustes Analysis. The methods are briefly introduced and used to address a recurring problem in sensory and consumer profiling research: assessors differ systematically in their judgements and in their interpretation and use of attributes.

Chapter 2

Assessing Panel Consonance[1]

In conventional sensory profiling studies where all assessors use the same attributes, it cannot be tacitly assumed that all attributes are understood equally well by all assessors. The lack of agreement can easily be shown by applying, e.g., a Generalised Procrustes Analysis to the data (see also Chapter 3). Even when the panel has been trained and uses a set of attributes with which they have experience, different use and interpretation of attributes still occurs. This lack of consistency of attribute-use is not unique to sensory and consumer data, though the lack of clear terms to describe odour sensations may be especially pernicious. In research where the agreement about one particular concept is desired, the same problem is encountered. The search by Spearman for "General Intelligence" in the early 20th century is an example where a -unidimensional- definition was sought. The lack of unidimensionality eventually resulted in the development of methods like Factor Analysis.

---

[1] The paper is printed in *Food Quality and Preference* (Dijksterhuis 1995).
Willem Heiser is thanked for comments on an earlier version of this paper. One of the reviewers of the paper for *Food Quality and Preference* is thanked for pointing out the problem of negative correlations.

The application of PCA in this chapter is to illustrate the (lack of) unidimensionality of the attributes assessors in sensory or consumer panels use. For each attribute a matrix with the scores the assessors gave the products is analysed. The method is based on Q-mode Principal Component Analysis, i.e., the assessor inter-correlation matrix is decomposed. When all assessors use an attribute in the same way this will give a large first eigenvalue. From the eigenvalues, or equivalently, the percentages of Variance Accounted For (VAF), a Unidimensionality- or *Consonance* index is computed. This index can be used to compare different attributes, or different stages of the training process of a panel. By plotting the loadings, representing assessors here, the method also enables easy visual inspection of the consistency of use of the attributes. Assessors that use a certain attribute differently can be identified. They may need further training, after which their use of the attribute can be inspected again.

The plots of the objects can be used to find products that may be responsible for the dissonant use of a certain attribute by one or more assessors. When products are presented in replications, these plots provide a rapid check on the validity of the result. Replicated products should lie very close in the object score plots.

The method assumes that all assessors use nominally the same attributes, so conventional profiling data is assumed; the method cannot be used with Free Choice Profiling data.

Chapter 3

Interpreting Generalised Procrustes Analysis "Analysis of Variance" Tables[2]

This chapter introduces "Projecting Procrustes Analysis" (GPPA, Peay 1988) as an alternative to the "classic" Procrustes Analysis of Gower (GPA, 1975) for the analysis of Conventional Profiling- and FCP-data. The two methods differ in some respects, the basic difference being the multidimensional space in which the configurations are matched. Figure 1 in Chapter 3 tries to explain the differences in terms of the partitioning of the total variance into four parts:

$$V_{total} = V_{in} + V_{out} + V_{within} + V_{projection} \tag{1}$$

(see also formula (1) in Chapter 3). A brief exposition may be appropriate here. In classic GPA, for each assessors' set $\mathbf{X}_k$, a rotation $\mathbf{H}_k$ is found that minimises the squared distances between the corresponding objects in the K sets. This rotation takes place in the space with the highest possible dimensionality. After the rotation there is nothing lost yet, all variance is still contained in this high-dimensional space. The result is K rotated matrices $\mathbf{Y}_k = \mathbf{X}_k\mathbf{H}_k$. The N rows, say $\mathbf{y}_{ki}$, with $k=1,\ldots,K$ and $i=1,\ldots,N$, of each of these matrices define the positions of KN objects in the aforementioned high-dimensional space. Because the corresponding points in the sets $\mathbf{Y}_k$ are as close together as is possible under rotations, the $\mathbf{Y}_k$ are averaged. The average space $\mathbf{Y}$ is called the *GPA Group Average*. This space also goes under the name *Consensus*, but, though introduced by Gower (1975), later, in Dijksterhuis and Gower (1991/2, §2.5, p. 72) it was noted that he has since avoided the term. The Group Average space $\mathbf{Y}$ contains the coordinates of the N objects, in the original high-dimensional space. To present this space in a lower number of dimensions, e.g. in the two dimensions of a graphical plot, a Principal Component Analysis is carried out on $\mathbf{Y}$. The configuration of Group Average points can now be inspected in a low-dimensional representation. The original NK points, i.e. the N objects for each individual assessor, can also be projected onto this space.

Two stages of the process induce a loss of variance:

1. the averaging of the $\mathbf{Y}_k$ to $\mathbf{Y}$
2. the projection of $\mathbf{Y}$ onto a low dimensional space by means of PCA

---

[2] The paper was published as an article in *Food Quality and Preference* (Dijksterhuis and Punter 1990).

The loss associated with 1 is called V_{within} in Chapter 3, the loss associated with 2 is called V_{out}. The criterion optimised by the GPA method is that the amount of variance V_{within} is as small as possible.

The GPPA method of Peay (1988) uses another space in which the optimal match of the configurations is defined. In addition to the rotations, a projection onto a low-dimensional space is performed by, say, $\mathbf{P}_k$. This rotation/projection is such that the variance contained in the low dimensional space is maximised. This variance is coined the *consensus*-criterion by Peay (1988). After the rotation/projection, the $\mathbf{Y}_k = \mathbf{X}_k\mathbf{P}_k$ are averaged to make the group average $\mathbf{Y}$. Note that this $\mathbf{Y}$ contains the coordinates of the N objects in the *low*-dimensional space. The configuration of the N points can now be inspected directly in the space $\mathbf{Y}$. In this process there are again two stages at which variance is lost:

1. the averaging of the $\mathbf{Y}_k$ to $\mathbf{Y}$
2. the projection of the $\mathbf{X}_k$ onto low dimensional spaces $\mathbf{Y}_k$ by means of $\mathbf{P}_k$

The loss associated with 1. is comparable to that in the GPA method, coined V_{within} in the chapter. The loss associated with 2. is different from the loss V_{out} in GPA. Note that the rotational part of $\mathbf{P}_k$ induces no loss; only the projection induces loss. This loss is called $V_{projection}$ in Chapter 3.

The variance remaining after the two losses are subtracted is called V_{in} for the GPA methods and $V_{consensus}$ for the GPPA method. More on the differences between GPA and GPPA can be found in Dijksterhuis and Gower (1991/2).

In the following chapter, the GPPA method is illustrated using a Conventional and a Free Choice Profiling data set. For both examples, the results for the products, the judges and the dimensions are shown. The distribution of the variances over the products and over the judges is shown in bar-charts, and the configurations of the products are plotted. To study the dimensionality of the solution, the percentages of the Variance Accounted For (VAF) are presented in a scree graph. Furthermore, it is suggested to express the VAF relative to the total variance in the data. This relative measure is easy to interpret and enables comparison of different Procrustes Analyses.

The use of bar-charts (Figures 4 and 8 in the chapter), and stacked bar-charts (Figures 2 and 6 in the chapter) for the representation of the variances, may be liable to discussion. This point is returned to in the Concluding Remarks which also contain some remarks about the interpretation of the Scree-graphs in the chapter (Figures 5 and 9).

CHAPTER 2

Assessing Panel Consonance

Garmt Dijksterhuis

(Originally published in *Food Quality and Preference*, 6, Dijksterhuis, G.B., Assessing panel consonance, 7-14, Copyright (1995), with kind permission from Elsevier Science Ltd, The Boulevard, Langford Lane, Kidlington OX5 1GB, UK)

CHAPTER 2

Assessing Panel Consonance

2.1 Introduction

In sensory and consumer research, a panel of assessors is often used to study properties of certain products, e.g. food-products. In the case presented here, the subjects give scores on attributes that refer to taste- and smell-perceptions. These attributes are often hard to define and it is well known that different subjects attach different meaning to the same attribute. This occurs even when the panel has been trained and uses a set of attributes with which they have experience. In Conventional Profiling, or other methods where all assessors use the same set of attributes, it can be useful to check for the different use of attributes. With Free Choice Profiling (Arnold and Williams 1985) this can be done by means of special Multivariate methods as Generalised Procrustes Analysis (GPA; Gower 1975, Dijksterhuis and Gower 1991/2) or Generalised Canonical Analysis (van der Burg 1988, van der Burg and Dijksterhuis 1989).

In this paper, another method is suggested to study the different use of the attributes by the assessors in a panel. It provides a check on the consistency, coined "consonance," of use of attributes by each individual assessor. The method assumes that all assessors use nominally the same attributes, so conventional profiling data is assumed, and the method cannot be used with Free Choice Profiling data.

Recently Schlich (1993a) published a method for an easy graphical representation of assessor performances. This method, coined GRAPES, is based on Analysis of Variance. Another method, also based on Analysis of Variance, was proposed by Næs and Solheim (1991). The plots provided by their method may become hard to interpret when large numbers of assessors and attributes are analysed.

The main objective of the *consonance*-method proposed in this paper is to provide the researcher with an easy graphical method to evaluate assessor's performance.

2.2 Data Structure

It is assumed that the sensory or consumer panel consists of K assessors who judge N products, using M attributes. The data from a such a panel can be arranged in a three-way data matrix $\mathbf{X}$ of order $(N \times M \times K)$ with elements x_{ijk} $(i=1,\ldots,N, j=1,\ldots,M, k=1,\ldots,K)$. An obvious way of analysing such a matrix is by first averaging over the assessors k, resulting in an $(N \times M)$ matrix with averages

$$x_{ij\cdot}=K^{-1}\sum_{k=1}^{K}x_{ijk} \tag{1}$$

which is subsequently analysed by Principal Component Analysis. It is then implicitly assumed that the M attributes are commensurate. However this assumption is often not met. The consonance-method proposed below provides an easy check on this assumption.

Usually the three-way data matrix $\mathbf{X}$ is thought to consist of K matrices $\mathbf{X}_k$, $k=1,\ldots,K$, of order $(N \times M)$, one for each assessor. Of course other slices of the three-way matrix can be taken, e.g. matrices $\mathbf{X}_j$, $j=1,\ldots,M$, of order $(N \times K)$ (see Figure 1).

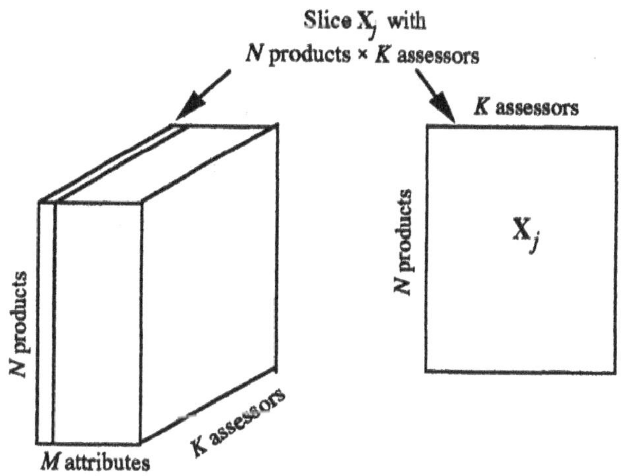

Figure 1 Three-way matrix with an $(N \times K)$ slice X_j.

The columns of the matrices $\mathbf{X}_j$ contain the same attribute j for each of the K assessors. Ideally these attributes, seen as variables, span a one-dimensional space. In that case the K assessors use the attributes similarly. In practice, the result will not be one-dimensional, which is the reason to apply an individual-difference technique, such as GPA, to the data matrix **X**. When the results of the above PCA show a significant deviation from unidimensionality, it's better not to average the data, and the attributes had rather be seen as interpreted and used differently by the assessors. Some assessors may need further training and/or some attributes better defining.

When the variables of the data matrix **X** are attributes that measure "preference," the consonance method is akin to MDPREF (Carroll 1972).

2.3 Method

Assuming standardised columns of $\mathbf{X}_j$, the matrix $\mathbf{X}'_j\mathbf{X}_j$ contains inter-assessor correlations which give an indication of the consistency of use of the jth attribute, but an underlying dimensional structure cannot be seen from the correlation matrix. A unidimensional result is signalled by a large first Eigenvalue, or "Variance Accounted For" value from the PCA applied to the slice $\mathbf{X}_j$ of the data matrix. Because a totally unidimensional result will hardly ever occur in practice, it can be useful to study the deviation from unidimensionality. One way of doing this is by constructing a two-dimensional representation of the N products. Such a representation can be obtained by plotting the object scores from the PCA, i.e. the first two columns of $\mathbf{Y}$ (this will be written $\mathbf{Y}^{[2]}$ from now on) where $\mathbf{Y} = \mathbf{XQ\Lambda}^{-1}$, $\mathbf{Q}$ is the matrix with Eigenvectors and $\mathbf{\Lambda}^2$ the matrix of Eigenvalues from the PCA of $\mathbf{X}$, which can be written as an Eigenvalue decomposition:

$$\mathbf{X}_j'\mathbf{X}_j = \mathbf{Q\Lambda}^2\mathbf{Q}' \tag{2}$$

$\mathbf{\Lambda}^2$ is a diagonal matrix which contains the Eigenvalues λ^2 in decreasing order along its diagonal. Since in this application the interest is mainly in the differences between the *assessors* and not between the *products*, a different plot may be more useful.

Note that the columns of the matrices $\mathbf{X}_j$ represent the K assessors. Plotting the component loadings, i.e. the row-points of $(\mathbf{Q\Lambda})^{[2]}$, gives positions for the assessors in a two dimensional space. This is the same space that the objects scores from $\mathbf{Y}^{[2]}$ lay in. These loadings are the correlations of the rows of the $\mathbf{X}_j$, i.e. the assessors, with the principal components.

The method as presented here is a linear method, assuming numerical (interval at least) scores. It can easily be extended into a non-linear method for the analysis of nominal, ordinal, numerical, or mixed data, by applying Princals (Gifi 1990) instead of ordinary PCA as presented here.

A perfect panel will consist of assessors that use all the attributes in more or less the same way. This will result in almost one-dimensional solutions of the PCA's on the matrices $\mathbf{X}_j$, signalled by a very high first Eigenvalue λ^2_1. Higher dimensions can then be regarded as error or noise. One way of comparing the first with higher dimensions is by looking at the ratio C (see consonance for a particular attribute) (C for *Consonance*):

$$C = \lambda_1^2 \left[\sum_{r=2}^{K} \lambda_r^2 \right]^{-1} \tag{3}$$

Of course the VAF values can be used instead of the Eigenvalues, and this does not change C.

A disadvantage of C may be that high negative correlations also contribute to a relatively high first eigenvalue, hence a high C. This may lead to an overestimation of the consonance when looking at the value of C only. It is important to inspect the assessor-plots in addition to the C-values. Negative correlations are clearly visible in the plots. On the other hand, negative correlations "only" reflect a reversed use of scale by an assessor. The attribute in question is not "different" in the sense that low-correlating attributes are different. It's a matter of discussion whether one wishes to call reversed attributes consonant.

Note that C looks similar to other test-statistics such as Student's t or Fisher's F-value:

$$C = \frac{\text{signal variance}}{\text{noise or error variance}} \approx \frac{\text{between variance}}{\text{within variance}} \tag{4}$$

It would be interesting to be able to use some distributional properties of C. When the nominator and denominator from (4) can be assumed to follow a distribution like a variance in normal samples, C will follow an F-distribution, but more research is needed here.

2.4 Examples

A number of examples are presented to illustrate the method. The examples are taken from different conventional sensory profiling experiments:

- cheese profiling data
- pea profiling data
- steak profiling data

2.4.1 Cheese Profiling Data

The cheese profiling data set is part of a study of the sensory properties of hard cheeses in which a Scottish and Norwegian panel are compared (Hirst *et al.* 1993). Here, only the data of the Norwegian panel were used.[1] In this study 10 assessors scored 12 different cheeses twice using 19 attributes. The attributes are divided into odour, flavour and texture attributes, but no reference to this division is made in the analyses. The replicated administration of the 12 cheeses results in 24 objects in the data matrix, which were analysed as if they were different products. Due to missing values, one of the attributes was removed. The remaining 18 attribute-data matrices $\mathbf{X}_j$ are of order (24×10).

---

[1] The data were kindly made available by Tormod Næs of Matforsk, Ås, Norway.

Eighteen separate PCA's were performed. The percentages of variance accounted for (%VAF) by the first four dimensions are shown in Figure 2. Each line represents the results of a PCA on one of the attribute-matrices $\mathbf{X}_j$, $j=1,\ldots,18$.

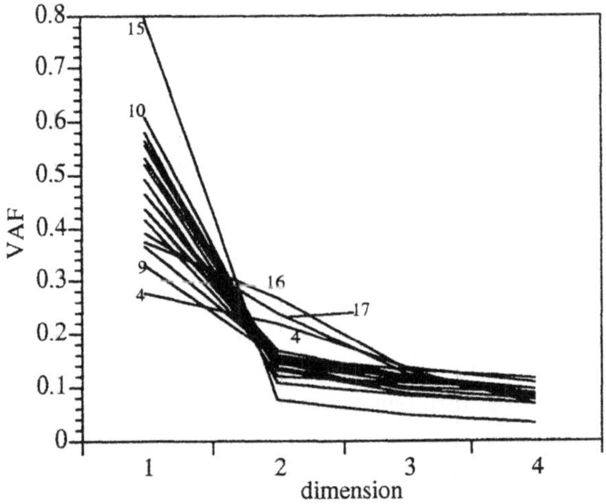

Figure 2 Proportion VAF per dimension of the separate PCA's of the 18 attribute matrices $\mathbf{X}_j$ of the cheese data (the numbers in the plot are attribute-numbers j).

It can be seen from Figure 2 that attribute 15 is closest to unidimensionality. Four attributes (4, 9, 16 and 17) can be seen to deviate most from unidimensionality. The plot of the *C-ratios* clearly indicates the same (Figure 3), but more clearly shows that attribute 9 is not unidimensional. In Figure 2 the attributes 4, 16 and 17 clearly stand out as not unidimensional because they have a large VAF value for their second dimension. These attributes are two-dimensional rather than one-dimensional. Attribute 9 does not have this two-dimensional nature, but the VAF values are spread more evenly among the second and higher dimensions. This is better illustrated in Figure 3, by means of a low value of C_9, than in Figure 2.

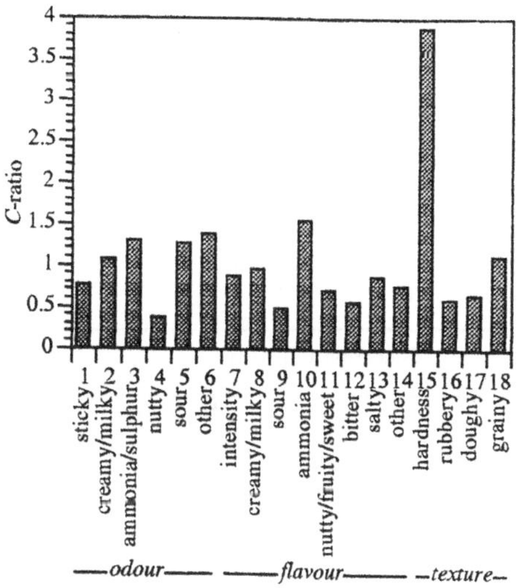

Figure 3 *C*-ratios of the attributes in the cheese data set.

Attribute 15 is far more unidimensional than the other attributes, and attribute 4 has the lowest consonance. In order to find out whether some assessors were responsible for this, the loadings of the PCA's of the corresponding data matrices $\mathbf{X}_4$, $\mathbf{X}_{16}$ and $\mathbf{X}_{17}$ and of $\mathbf{X}_{15}$ are plotted and inspected. Figure 4 presents these plots for the attributes 4, 16 and 17 and for attribute 15.

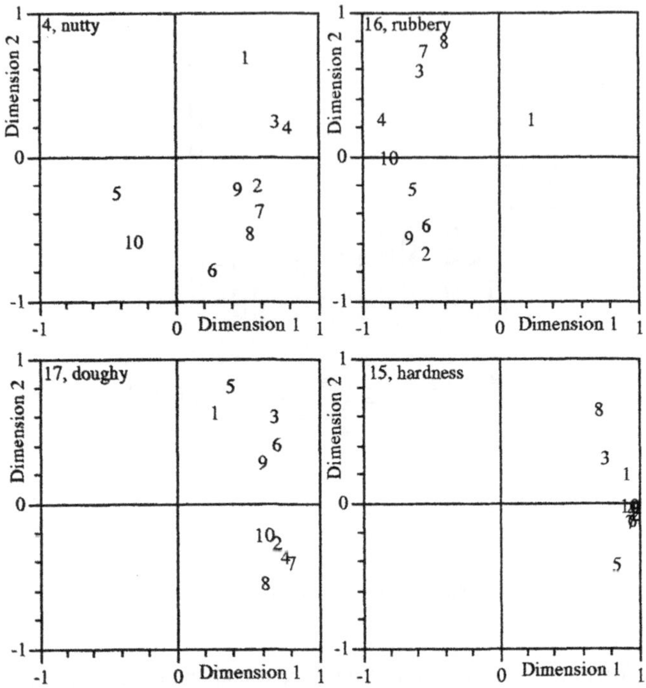

Figure 4 Plot of the assessors from the PCA's of the most unidimensional attribute (no. 15) and the three least unidimensional attributes (no. 4, 16, 17) in the cheese data set.

In Figure 4 assessor 1 can be identified as partly responsible for the deviation in scores for the attributes 4 and 16. Two groups of assessors can be seen in the plot for the other less unidimensional attribute (no. 17). One group consists of the assessors 1, 3, 5, 6 and 9, the other group of the numbers 2, 4, 7, 8 and 10. This illustrates the potential of this method to identify segments of assessors with respect to certain attributes. Especially in consumer studies with a large number of respondents, this could prove useful.

When replicate information about some products is present in the data, the plotting of the object scores from the PCA gives information about the consistency of scoring of the replicates with respect to each attribute separately. In this example, each cheese sample was judged twice, and in the analysis the two scores were treated as if they came from different products. As a result the data matrix contained N rows, while only ½N different product samples were involved. It is interesting to compare the position of the members from a pair of identical samples. In order to do this comparison the corresponding points were connected with each other in Figure 5.

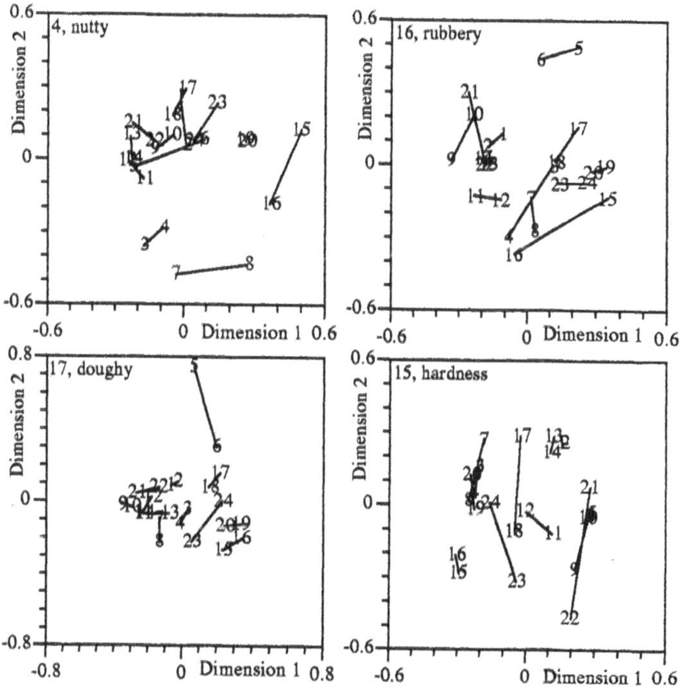

Figure 5 The *N* cheese sample-points in the objects space of the PCA on the attribute matrices for the most unidimensional attribute (no. 15) and the three least unidimensional attributes (no. 4, 16, 17). The elements of the pairs of identical products are connected by a line.

From Figure 5 can be seen that there seems to be no clear systematic difference between the two samples of each pair. This is different from what is sometimes found in studies where the replicated samples were judged in a different session (see e.g. Dijksterhuis 1990/1993). In the cheese study, care was taken to avoid effects of order of presentation by using an appropriate test plan (MacFie *et al.* 1989, see also Schlich 1993b).

2.4.2 *Pea Profiling Pata*

The pea data set comes from a study of the sensory properties of peas[2] (Kjølstad *et al.* 1990, Næs and Kowalski 1989). In this study each pea sample was presented twice. In contrast with the previous example on cheese, now the two presentations were analysed separately. This gives two data sets, one containing the first presentations and one the second presentations of each pea-sample. Both data sets consist of the scores of 10 assessors who scored 60 pea-samples using 6 attributes. The attribute-data matrices $\mathbf{X}_j$ are of order (60×10) and 6 separate PCA's were performed for each of the two presentations. The percentages variance accounted for (%VAF) for the first four dimensions are shown in Figure 6.

[2] The data were kindly made available by Tormod Næs of Matforsk, Ås, Norway.

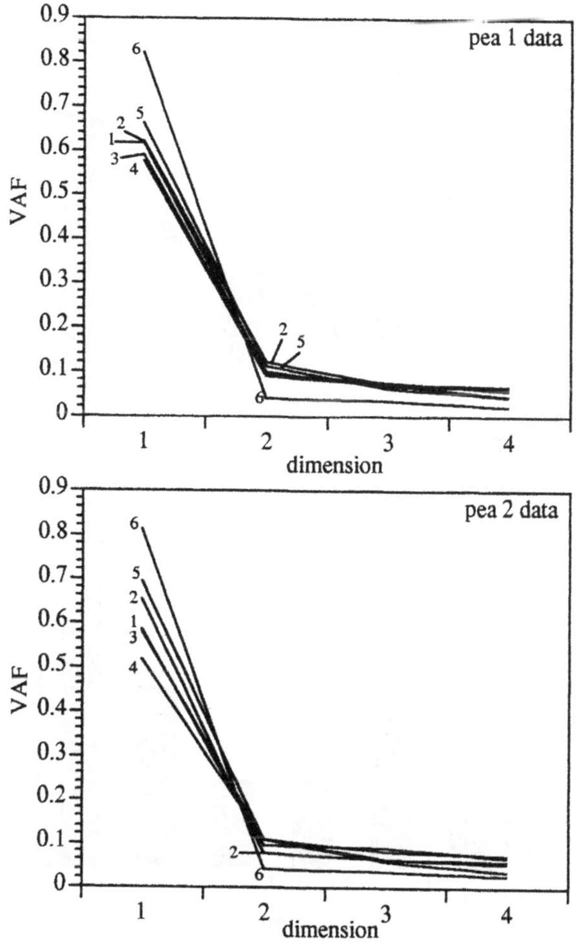

Figure 6 Proportion VAF per dimension of the separate PCA's of the 6 attribute matrices of the two pea-sample presentations ("pea 1 data" and "pea 2 data").

It can be seen from Figure 6 that all attributes are rather unidimensional, the differences are much smaller than in the previous example on cheeses, but there more attributes were involved. Attribute 6 has the highest consonance. No attributes show large deviations from unidimensionality. The plot of the *C*-ratios supports the same conclusion (Figure 7).

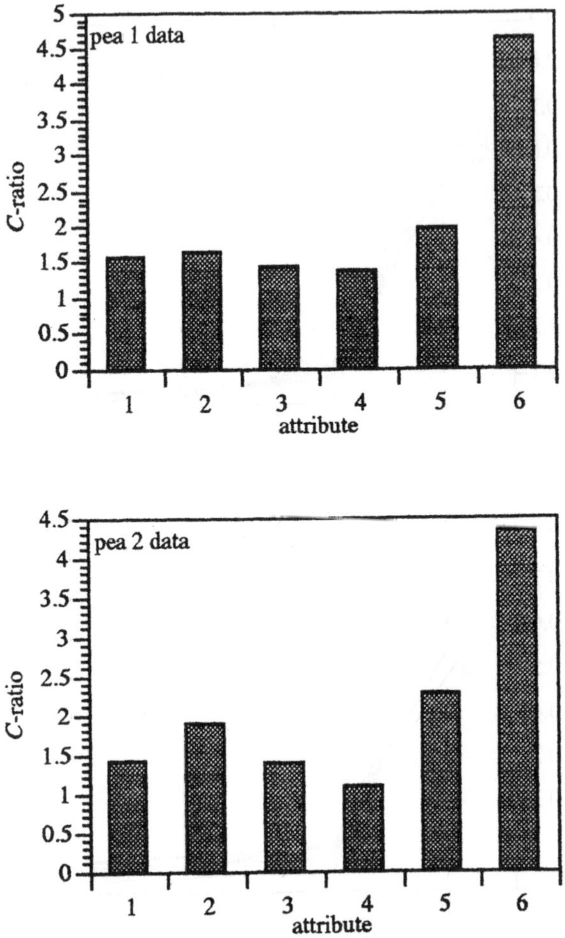

Figure 7 *C*-ratios of the attributes in the two pea data sets.

Attribute 6 is more unidimensional than the other attributes. The two presentations of the peas resulted in about the same consonance results for the 6 attributes. Attribute 4 is the less unidimensional in both pea-sample presentations. Figure 8 presents the assessor plots for attribute 4 and 6.

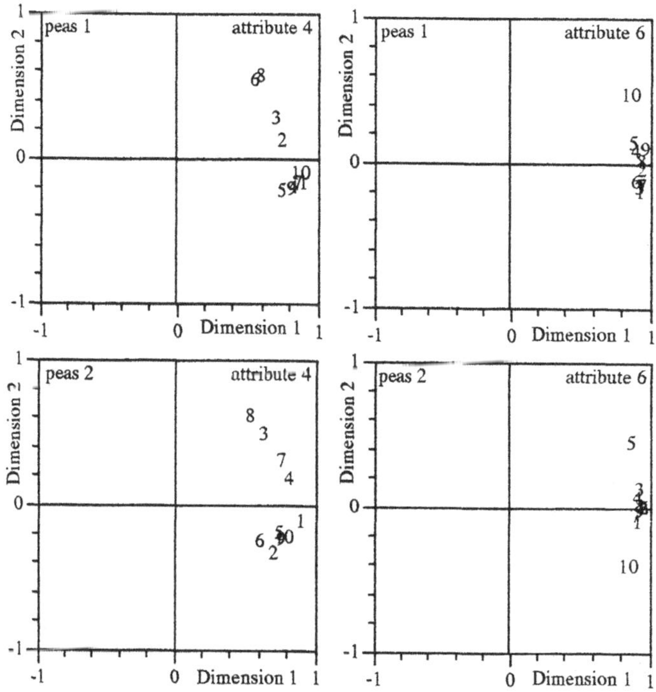

Figure 8 Plot of the assessors from the PCA's of the most unidimensional attribute (no. 6) and the least unidimensional attribute (no. 4) for both parts of the pea data set ("peas 1" and "peas 2").

Figure 8 shows clearly that for both pea data sets, attribute 6 is unidimensional, and attribute 4 is less so. The loadings the attributes of the 10 assessors received are almost equal for the first dimension, but they differ for the second. Assessors 3 and 8 seem to be deviating the most for attribute 4, and for attribute 6 the assessors 5 and 10 apparently scored different from the others.

2.4.3 *Steak Profiling Data*

The steak data study is reported by Rousset *et al.* (1992) and Schlich (1993b).[3] The data set contains scores of 11 assessors on 16 steaks using 13 attributes. The attribute-data matrices $\mathbf{X}_j$ are of order (16×11) and 13 separate PCA's were performed. The percentages of variance accounted for (%VAF) for the first four dimensions are shown in Figure 9.

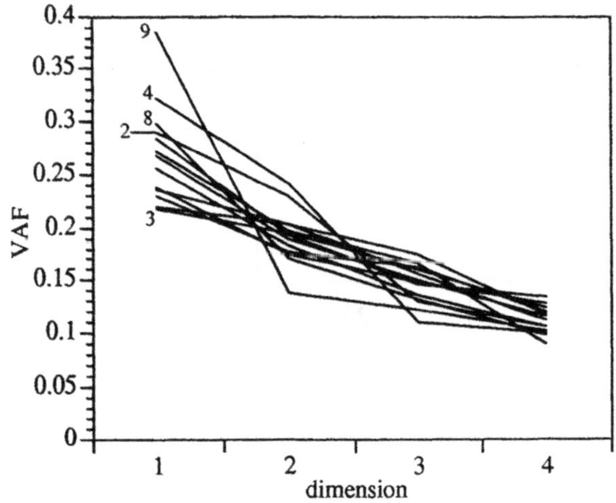

Figure 9 Proportion VAF per dimension of the separate PCA's of the 13 attribute matrices of the steak data set.

Comparison of the screes in Figure 9 with the screes of the previous examples shows that the former are much flatter, implying less overall consonance than the latter. This probably explains the trouble in interpreting the consonance of attribute 4. From Figure 9 it shows that attribute 4 and 2 are the least unidimensional, attribute 9 is the most unidimensional. Though attribute 4 has the second highest VAF value in the first dimension, it has the highest VAF value in the second dimension, this attribute is rather two- than one-dimensional. The *C*-ratios in Figure 10 however show attribute 4 as the second most

[3] The data were kindly made available by Pascal Schlich, INRA, Dijon, France.

consonant attribute and indicate that attribute 5 is another low-consonance attribute. This ambiguity is also exemplified by the relatively low *C*-ratios presented in Figure 10.

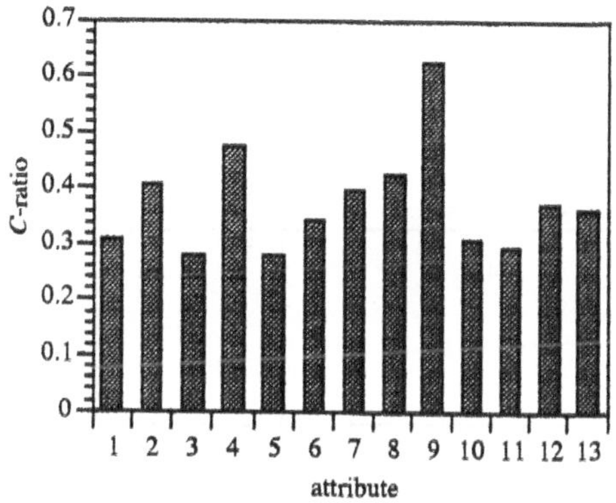

Figure 10 *C*-ratios of the attributes in the steak data set.

The scale of the ratios is much lower than in the previous examples, which means that the quality of the steak-panel, in terms of consonance of the attributes, is lower than that of the cheese- and pea-panel. Table 1 presents some statistics of the *C*-values, showing that there is something different in the steak-data compared to the other data sets.

Table 1. Mean, minimum, maximum and range for the *C*-ratios from the four data sets

data set	mean	minimum	maximum	range
cheese	1.0680	0.3847	3.8903	3.5056
peas 1	2.1103	1.3586	4.6515	3.2929
peas 2	2.0600	1.0794	4.3252	3.2503
steak	0.3758	0.2797	0.6260	0.3463

This can also be seen in the plots in Figure 11. This figure will enable the identification of assessors that are responsible for the lack of consonance.

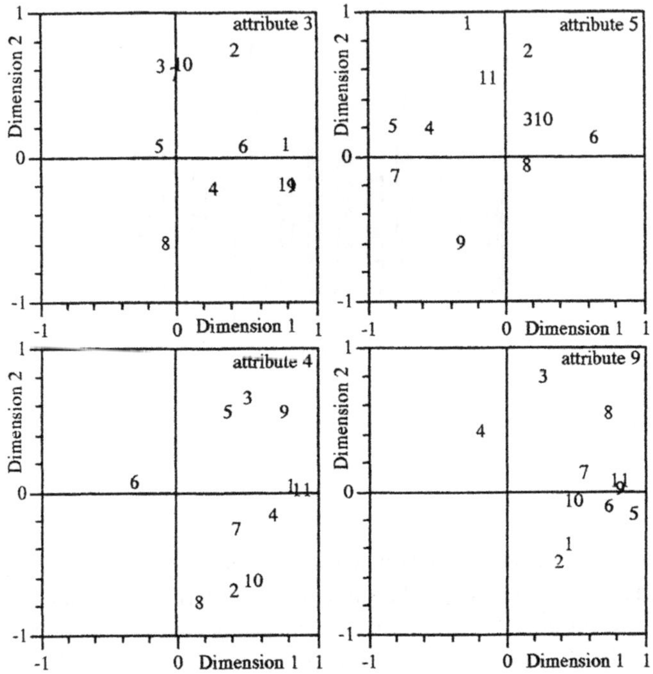

Figure 11 Plot of the assessors for the attributes 3, 5 (low consonance), 4 and 9 (high consonance) from the steak-data.

Figure 11 clearly shows that the general unidimensionality of the attributes in the steak data set is lower than in the other data sets analysed. The figures clearly show more variance in the first dimensions, immediately visible by a cloud of points much wider than in the previous examples. For attribute 4 assessor 6 is different from the other assessors; for attribute 9 assessors 3, 4 and 8 deviate from the other assessors. The two low-unidimensionality attributes (no. 3 and 5) show no clear structure in the positions of the assessors. In this case a Generalised Procrustes Analysis could be used to match these configurations of the assessors to look for underlying similarities of the configurations. This is not done here, but would be an interesting extension to the method.

2.5 Conclusion

The method proposed enables a fast assessment of the homogeneity, or "consonance," of scoring between the assessors in a sensory profiling or consumer profiling panel. The plots of the *C*-ratios and the scree graphs of the proportions Variance Accounted For (VAF) can be used to identify attributes which may have different meanings for the assessors. The plots of the assessors provide an easy graphical check on the homogeneity of the panel with respect to each attribute. In addition, assessors can be identified as outliers and segmentations of the panel can be found.

As a method of finding attributes that are used consistently throughout the panel, "consonant attributes," the *C*-ratios can be inspected and attributes below a certain value may be discarded. It could then be decided that the panel, or some deviant assessors, require further training with respect to these attributes.

CHAPTER 3

Interpreting Generalised Procrustes Analysis "Analysis of Variance" Tables

3.1 Introduction

The field of Generalised Procrustes methods is still lively and a new method was published by Peay (1988). This method is implemented in the PROCRUSTES-PC v2.0 program (OP&P 1989, Dijksterhuis and van Buuren 1989). Because there are now a number of different GPA software packages using different methods, it may be useful to be able to compare the results obtained by different analyses and by different software. It is important to realise that these methods do not differ merely in different computer implementations or algorithms, but fit different types of configurations, and hence should be distinguished by different names. Here, 'GPA' (Generalised Procrustes Analysis) is used for the methods based on Gower's 1975 article. Peay's method is referred to as GPPA (Generalised Projection Procrustes Analysis).

(Originally published in *Food Quality and Preference*, 2, Dijksterhuis, G.B. and Punter, P.H., Interpreting Generalised Procrustes Analysis "Analysis of Variance" Tables, 255-265, Copyright (1990), with kind permission from Elsevier Science Ltd, The Boulevard, Langford Lane, Kidlington OX5 1GB, UK)

In this article a measure relative to the total amount of variance in the data is proposed. This measure is easier to interpret and related to the V.A.F.-(Variance Accounted For) measure and the percentage 'explained' variance. A relative measure makes it easier to compare solutions from different Procrustes analyses made by different software. The 'Analysis of Variance' tables produced by the software can be standardised using the relative measure proposed in this article. These tables can be of great help in interpreting the solution of the analyses. They show the relation between the fit of the solution, the dimensionality, the agreement between the judges and the agreement about the products.

3.2 Two Different Procrustes Methods

Since its introduction (Green 1952, Hurley and Cattell 1962; Cliff 1966, Schönemann 1966) Procrustes Analysis has been in constant development. The early Procrustes Analysis was the so-called one-sided orthogonal Procrustes Analysis in which one configuration was rotated to another configuration (the target). Later extensions included a scaling factor in addition to the rotation (Schönemann and Carroll 1970). The original method was generalised to include more than two configurations by Kristof and Wingersky (1971). Gower (1975) extended the method by including a scaling factor in the generalised case. The Gower article is the most cited one in the field of Generalised Procrustes Analysis and most GPA software is based on the Gower method. He also provided an 'Analysis of Variance' table with goodness of fit measures for the products and the judges. The Gower algorithm was improved by ten Berge (1977).

The characteristic of these GPA methods is that they perform a symmetric analysis. This means that it is necessary for all individual sets to contain the same number of attributes, i.e., each judge must use the same number of attributes. Because Free Choice Profiling (Williams and Langron 1984) results in data that can differ in the number of attributes per judge, the number of attributes must be made the same for each judge. Usually this is done by appending attributes containing only zero's to the sets until all sets have the same number of columns. Next the sets, which all have the same size now, are matched by rotating and scaling them. This procedure minimises the Least Squares or Procrustes criterion, which measures the difference between the average- or 'consensus'-space and the individual spaces. When this difference is small enough, the spaces are averaged to produce the consensus-space. From this space the experimenter is usually interested in two or three principal components which are extracted by a regular Principal Component Analysis.

Ten Berge & Knol (1984) developed a method to rotate sets with possibly different numbers of attributes to a consensus configuration of low dimensionality. This method was extended by Peay (1988) to the GPPA method. Peay provided a scaling step and partitioned the total variance in the data into three distinct parts. The GPPA method of Peay differs from Gower's GPA approach in that it does not minimise the squared residuals but it maximises the variance of the low dimensional consensus-space obtained. In addition to the rotation and scaling of the sets, they are also projected onto a subspace of a dimensionality chosen by the experimenter. Often the consensus-space is chosen to be two or three-dimensional. The dimensionality of the resulting consensus-space does not need to be reduced any further by PCA because it's already of low dimensionality.

Although the two methods are different, in practice they produce almost visually identical results. Whether or not the empirically determined similarity of the methods is necessary is not clear and the exact differences and their theoretical and practical implications are the subject of further study.

3.3 Sums-of-squares in Generalised Procrustes Analysis

In Gower's 'Procrustes Analysis of Variance' (1975, Table 4) the total variance is distributed over the N products and the K judges to show the fit of the products and the agreement of the judges with the consensus solution. In this table, the reported variances between and within judges seem to be swapped. The score by judge k on attribute j about product i is represented by $\mathbf{x}_{kij}$.

The total variance can be partitioned as follows:

$$V_{\text{Total}} = V_{\text{In}} + V_{\text{Out}} + V_{\text{Within}} + V_{\text{Projection}} \qquad (1)$$

in which V_{Total} is the total variance contained in the data, or is the sum-of-squares over all K sets, N products and M_k attributes:

$$V_{\text{total}} = \sum_{k=1}^{K} \sum_{i=1}^{N} \sum_{j=1}^{M_k} \mathbf{x}_{kij}^2 \qquad (2)$$

The parts V_{In} and V_{Out} together constitute the consensus variance:

$$V_{Consensus} = V_{In} + V_{Out} \tag{3}$$

being the criterion maximised by the Peay method. V_{In} is the part 'explained' by the first Q dimensions of the consensus-space, and V_{Out} is the part left unexplained. This is the part associated with the higher dimensions of the consensus-space. Since in this article only results from the Peay method will be discussed, it is not useful to pay much attention to V_{Out} (V_{Out} is only nonzero with the Gower method as can be seen in Figure 1).

The within- and projection-variance are residuals stemming from two different sources. V_{Within} is the part lost in averaging the obtained individual spaces to the consensus-space. V_{Within} constitutes the Least Squares or Procrustes criterion minimized by the Gower method. $V_{Projection}$ is the variance lost in the process of projecting data onto the subspaces.

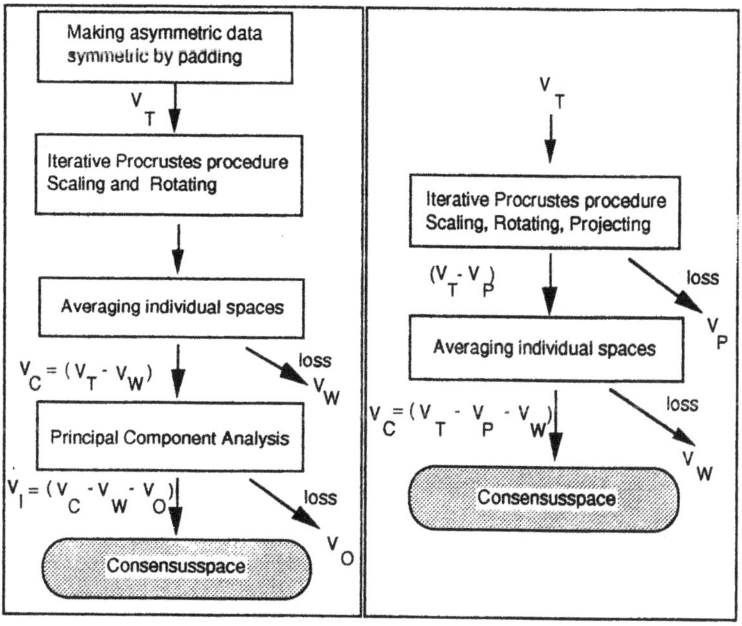

Figure 1 The two step process of Generalised Procrustes Analysis with the associated variances V_{Total}, $V_{Projection}$, $V_{Consensus}$ and V_{Within} (Respectively V_T, V_P, V_C, and V_W).
Left: According to the Gower (1975) method. Right: According to the Peay (1988) method.

Figure 1 reflects the two step nature of Generalised Procrustes Analysis. The right part of Figure 1 shows the two steps with the corresponding amounts of variance using the Peay method. The first step is the iterative procedure in which the individual configurations are matched using the admissible transformations of rotation and scaling while at the same time projecting them onto lower-dimensional subspaces. The loss associated with this step is $V_{Projection}$. The second step is the simple averaging of the individual spaces obtained by the first step. The loss associated with this step is V_{Within}. What is left after these two steps is $V_{Consensus}$, the consensus variance. This implies that the part of the total variance left after the first step (V_T-V_P) is distributed over V_{Within} and $V_{Consensus}$.

The left part of Figure 1 shows the situation with the Gower method. In this situation the role of $V_{Projection}$ is played by V_{Out} in the final Principal Component Analysis. The term $V_{Projection}$ does not appear in this method; it is zero since the Gower method does not project onto subspaces. On the other hand, V_{Out} does not appear in the Peay method because it does not perform a PCA on the full-dimensional consensus-space, as a result V_{Out} is zero with the Peay method.

3.4 Scaling the Total Variance

In both Procrustes methods, the total variance V_{Total} is scaled prior to the iterative procedure and remains constant (γ) throughout it. There are different ways to scale V_{Total}. Peay provides the most general formula, using Gower's scaling as a special case:

$$\gamma = (\max_k\{M_k\})^{-1} \sum_{k=1}^{K} M_k \tag{4}$$

γ is the total number of attributes divided by the maximum number of attributes. In the case of symmetric data or data made symmetric by the padding of zero-columns, γ amounts to the number of sets K. The exact value of the constant γ to which the total variance is scaled is not relevant to the Procrustes procedure and can in fact be chosen freely (Gower 1975, p. 37, Peay 1988, p. 205). However, the way the variance is scaled has implications for the way the 'Analysis of Variance' tables look. The GENSTAT (Arnold 1986) and SAS (Schlich 1989) macros present these tables in original units, i.e., the variances reported are not scaled. The PROCRUSTES-PC program scales according to Peay. Because of the differences in scaling it is not easy to compare the results from different programs. Even the results from one program are not directly comparable when the scaling according to Peay or Gower is used since this scaling depends on the total and maximum number of attributes (see formula 4).

It is possible, however, to compare the Analysis of Variance tables in an easy and 'natural' manner by reporting the results *relative* to the total amount of variance. In this way the Analysis of Variance results correspond to the familiar 'explained variance' or 'Variance Accounted For' (V.A.F.) concept. The reported Analysis of Variance results can always be expressed as proportions of the total rescaled variance by the user of the program after analysing the data. The developers of Generalised Procrustes software would be much more user-friendly if they were to include this in the next release of their programs. Version 2.1 of the PROCRUSTES-PC program has the possibility of presenting the variances as percentages VAF (OP&P 1990, Dijksterhuis and van

Buuren 1990). This can be easily accomplished by making the constant γ equal to 100 and causes no trouble with computer precision because 100 is a sufficiently small number.

3.5 Generalised Procrustes Analysis of a Conventional Profiling Experiment

The results presented here are from the data of 31 judges, using nine attributes to describe five kinds of cheese. It is a conventional profiling experiment which means that every judge uses the same attributes. The results are based on a two-dimensional analysis using the Peay GPPA method implemented in the PROCRUSTES-PC program. Table 1 shows the unscaled between, within and total variance.

Table 1 Between, within, total and rescaled total variance of the cheese data.

Between sets	187042.06
Within sets	539932.00
Total	726974.06
Rescaled	100

The 'Between sets' and the 'Within sets' variances are in original units. The sum (V_{Total}) is rescaled to $\gamma = 100$. As a result all other reported variances can be interpreted as percentages of the total variance. In the PROCRUSTES-PC v2.1 program, the rescaling of the total variance is optional. Either is rescaled according to

$$\gamma = \frac{\text{total number of attributes}}{\text{maximum number of attributes}} \tag{5}$$

(see also formula 4), to 100 as proposed in this paper, or it is not scaled at all. Relative variances can be calculated by dividing the values in the tables by the rescaled variance γ. This holds true regardless of how the total variance is scaled, so the results in original units from the GENSTAT or the SAS macro can easily be converted to relative measures.

The total variance V_{Total} is decomposed into four parts (see formula 1). Each of the parts can be partitioned over the units in the analysis, being the products and the subjects. As can be seen in Figure 1, $V_{Projection}$ is lost in the Procrustes procedure, and what is left is (V_{Total} - $V_{Projection}$) is shown in the tables. Though this variance is called 'Total' in the PROCRUSTES-PC program output, it is the variance left *after projecting*, it is not the same as V_T, which is the total variance *before* any Procrustes transformation is carried out. The variance left after projecting can be decomposed further into $V_{Consensus}$ and V_{Within}. Table 2 illustrates this using the data from the cheese experiment.

Table 2 Rescaled consensus, within, projection and total variance of the cheese data.

$V_{Consensus}$	V_{Within}	$V_{Projection}$	V_{Total}
67.013	15.186	17.801	100

$V_{Projection}$ is 17.8% of the total variance so 17.8% of the total variance was removed in projecting the individual configurations onto the two-dimensional individual spaces. Of the remaining 82.2%, a part is lost again as V_{Within} in the process of averaging the individual configuration to the consensus-space. In this case V_{Within} amounts to 15.2%. The consensus-space still accounts for 67% of the total variance, which is a respectable amount for two dimensions.

3.5.1 *Products*

Table 3 shows the percentages of the different variances associated with the five products from the cheese experiment.

Table 3 Percentage consensus and within-variance distributed over the five different kinds of cheese.

Products	$V_{\text{Consensus}}$	V_{Within}
1	11.538	2.877
2	9.327	3.605
3	0.447	3.819
4	22.349	2.723
5	23.352	2.162
Total	67.013	15.186

The 67% total consensus variance is distributed over the five products as shown in the column $V_{\text{Consensus}}$. Figure 2 presents the variances from Table 3 in graphical form. The lower, lighter part of the histogram represents the percentage consensus variance, the upper darker part the residual within variance of the corresponding products.

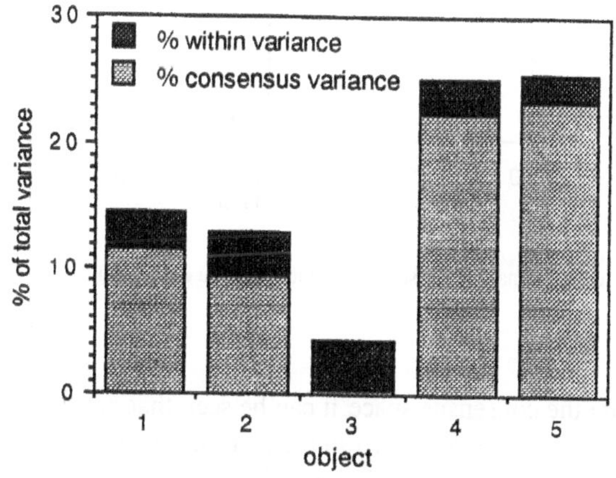

Figure 2 Percentages consensus- and within variance distributed over five different kinds of cheese.

The products 4 and 5 represent respectively 22.4% and 23.4% of the total variance in the consensus-space. The residual 15.2% within-variance is also distributed over the products, which can be seen in the column V_{Within} in the table. These variances are the lowest (respectively 2.2% and 2.7% of the total variance) for the products 4 and 5. This means there exists agreement between the judges about the position of these products. The products 1 and 2 are intermediate with reference to the consensus and residual variances. Product 3 only has 0.5% consensus variance associated with it, and the residual variance is the largest (3.8%) for this product. This means that in the process of projection to the two dimensional subspace almost all variance for product 3 is lost. This cheese explained the 5.5% consensus variance in a three-dimensional analysis. The consensus-space for the five products is shown in Figure 3.

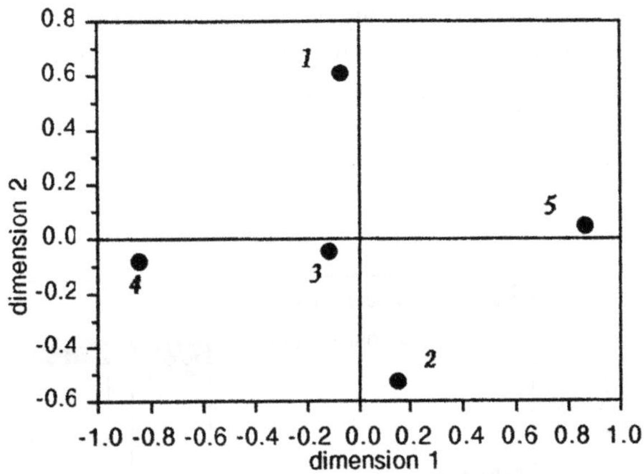

Figure 3 Two dimensional consensus-space with the five different kinds of cheese.

From the consensus-space it can be seen that product 3 lies close to the centre of the space. This results from the fact that cheese number three occupies a different position in each individual judge's space. The averaging of the individual spaces results in a central position of this product. It can be concluded that the product is assessed differently by the individual assessors.

3.5.2 Judges

Table 4 presents the distribution of the residual within-variance over the 31 judges.

Table 4 Percentages residual V_{Within} distributed over the 31 judges in the cheese experiment.

Judge	V_{Within}	Judge (ctd.)	V_{Within} (ctd.)	Judge (ctd.)	V_{Within} (ctd.)
1	0.464	12	0.825	23	0.182
2	0.278	13	0.447	24	0.487
3	0.687	14	0.607	25	0.499
4	0.437	15	0.675	26	0.387
5	0.455	16	0.640	27	0.541
6	0.340	17	0.402	28	0.416
7	0.497	18	0.270	29	0.266
8	0.433	19	1.061	30	0.540
9	0.599	20	0.532	31	0.238
10	0.559	21	0.532	-	-
11	0.354	22	0.536	-	-
-	-	-	-	total	15.186

As can be seen from Table 4, 15.2% (V_{Within}) of the total variance is distributed over the judges. Figure 4 gives a graphical representation of the distribution of V_{Within} over the 31 judges.

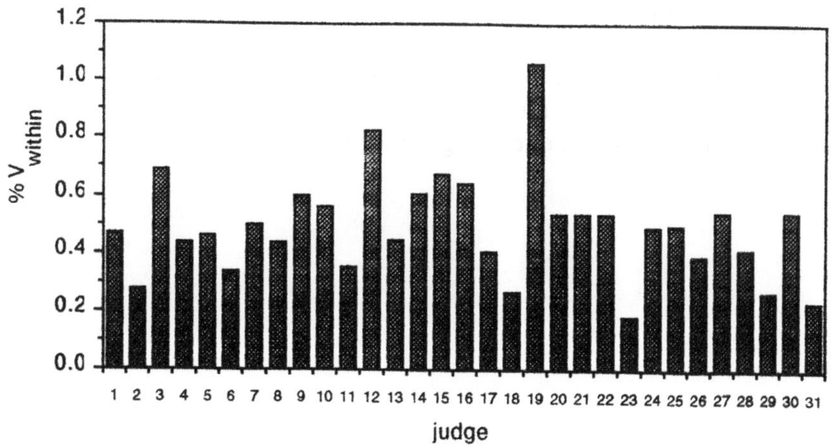

Figure 4 Distribution of the residual V_{Within} variance over the 31 judges.

If the value for V_{Within} is large for an individual judge it means that this judge does not agree with the consensus-space. Judges 3 (0.7%), 12 (0.8%), 15 (0.7%) and 19 (1.1%) are the ones with the larger amounts of residual variance.

3.5.3 Dimensions

Table 2 showed that the two dimensional consensus-space (Figure 3) explains 67% of the total variance. The distribution of the 67% over the two dimensions is shown in Table 5.

Table 5 Percentage consensus- and within- variance distributed over the two dimensions of the consensus-space.

Dimensions	$V_{Consensus}$	V_{Within}
1	46.650	9.374
2	20.363	5.812
total	67.013	15.186

From Table 5 it can be seen that the first dimension of the consensus-space explains 46.7% of the total variance and the second dimension explains 20.4%.

It can often be useful to know the distribution of the variance over all dimensions to help in choosing the dimensionality of the consensus-space. For this reason, a full-dimensional analysis of the cheese data (in this case a 4 dimensional analysis) was run. The percentages consensus variance explained by the four-dimensions of the resulting consensus-space are shown in Table 6.

Table 6 Percentage consensus variance per dimension and cumulative, contained in the four dimensions of the full-dimensional consensus-space of the cheese data.

Dimensions	$V_{Consensus}$	cum $V_{Consensus}$
1	45.609	45.609
2	19.318	64.927
3	13.849	78.776
4	6.518	85.294
Total	85.295	

From Table 6 it can be seen that four-dimensions explain 85%, and three-dimensions explain 79% of the total variance. The first two dimensions explain 65% instead of the 67% reported in Table 5. This is due to the Procrustes method used. Because the method includes optimal projection onto subspaces, the solutions are not *nested* as in the classical Procrustes methods. The classical GPA methods perform a PCA on a full-dimensional consensus-space to extract a sufficiently low number of principal axes (see Figure 1). Because in this case the dimensionality of the consensus-space stems always from the same PCA, the solutions of these GPA methods are nested.

Figure 5 shows a scree graph to assist in determining the dimensionality of the solution. The graph is based on the data from Table 6.

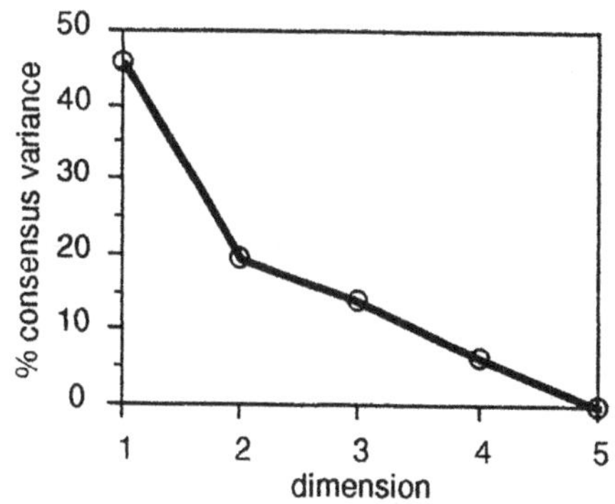

Figure 5 Scree graph of consensus variance per dimension for a full-dimensional solution of the cheese data.

From Figure 5 and Table 6 it can be inferred that a three-dimensional GPPA solution might be useful. To find out it also was computed and it explains 79.8% variance, which is a little more than the 78.8% from Table 6 due to the non-nestedness of the method. In fact cheese number three explained 5.5% in the three-dimensional solution and the residual within variance dropped just a little to 3.5%. It can be concluded that there does seem to exist some kind of agreement between the assessors about this cheese, but that it is not as clear as with the other cheeses. In order to find out more about this cheese, the correlations of the attributes with the consensus dimensions can be inspected; this is beyond the scope of this paper.

3.6 Generalised Procrustes Analysis of a Free Choice Profiling Experiment

The second series of results comes from a Generalised Procrustes Analysis of free-choice profiling data. Nine judges assessed eight different yoghurts, each using his/her own attributes. The number of attributes ranged from 8 to 17. The judges received little or no training. Below the results from a two-dimensional analysis are presented. Like in the previous example, the analysis was carried out to extract a two dimensional consensus-space with the PROCRUSTES-PC program.

Table 7 Between, within, total and rescaled total variance of the yoghurt data.

Between sets	131012.43
Within sets	437967.87
Total	568980.30
Rescaled	100

The total variance was scaled to 100 (see Table 7) just as with the cheese experiment above. Table 8 presents the different sources of rescaled variance, $V_{Consensus}$, V_{Within}, $V_{Projection}$ and V_{Total}.

Table 8 Rescaled consensus, within, projection and total variance of the yoghurt data.

$V_{Consensus}$	V_{Within}	$V_{Projection}$	V_{Total}
59.461	10.675	29.864	100

Comparison with Table 2 shows a considerable difference in $V_{Projection}$ (17% versus 30%). This is probably due to the fact that in this Free-Choice profiling experiment, the judges differed more in their assessments because they received less training.

After the Procrustes procedure (see Figure 1) the averaging of the individual spaces results in a 10.7% loss. The two-dimensional consensus-space accounts for 59.5% variance, which is 7.5% less than the conventional profiling experiment on cheese shown above.

3.6.1 Products

In Table 9 the distribution of the consensus- and within-variance over the 8 different yoghurts is shown.

Table 9 Percentages consensus- and within variances distributed over the 8 yoghurts.

Product	$V_{Consensus}$	V_{Within}
1	0.765	1.429
2	10.328	0.772
3	6.538	2.629
4	4.098	2.207
5	8.028	0.418
6	7.833	1.318
7	14.179	1.101
8	7.693	0.801
Total:	59.461	10.675

Figure 6 shows the data from Table 9 in graphical form (compare Figure 2). It shows that yoghurts number 7 and 2 fit well into the consensus-space compared with the other yoghurts. There must have been agreement between the judges about these yoghurts. This is not true for number 1, because its 'explained variance' $V_{Consensus}$, is very small. Numbers 3 and 4 have the largest within variances, presumably because the judges did not agree on these yoghurts.

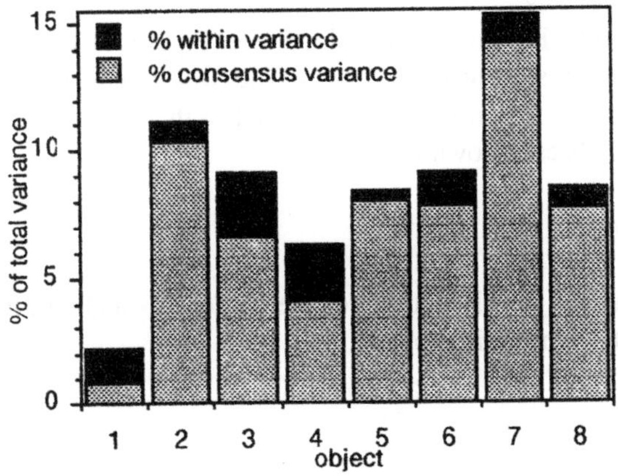

Figure 6 Percentages consensus- and within variance distributed over eight different kinds of yoghurt.

In Figure 7 the consensus-space is shown, and it can be seen that there are two pairs of yoghurts, numbers 2 and 6 and numbers 5 and 8, which lie very close. From Figure 6 it can be seen that V_{Within} and $V_{Consensus}$ are virtually the same for yoghurts 5 and 8. In fact these were duplicates, presented blindly to the subjects. It shows that these two yoghurts were judged almost identically.

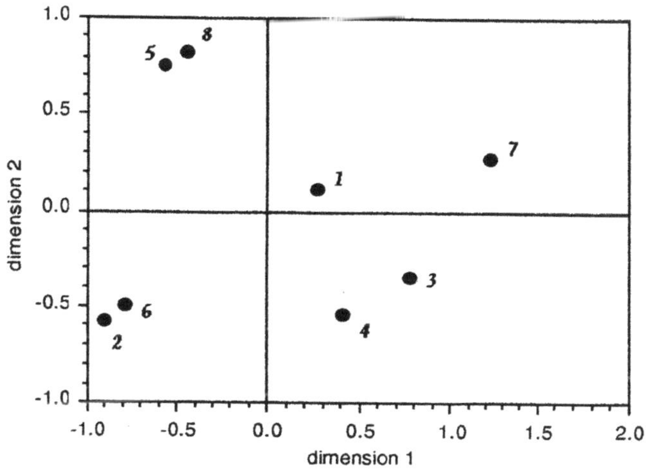

Figure 7 Two dimensional consensus-space with the eight kinds of yoghurt.

The other pair (2 and 6) turns out to consist of two variants from the same brand, number two the light (low fat content) and number 6 the normal one. Apparently the two variants were very much alike for the judges. Number one was shown by Figure 6 to be the yoghurt that has the least agreement between the judges, since it lies close to the centre of the consensus-space. It lies in a different position in the individual spaces for each judge.

3.6.2 *Judges*

Table 10 shows the distribution of the residual variance V_{Within} over the nine judges in the yoghurt experiment.

Table 10 Percentages within variance distributed over the nine judges in the yoghurt experiment.

Judge	V_{Within}
1	0.747
2	1.052
3	1.408
4	0.956
5	1.471
6	1.631
7	2.199
8	0.660
9	0.550
Total	10.675

Figure 8 shows a graphical representation of the amount of unexplained variance V_{Within} distributed over the nine judges. It contains the same information as Table 8 but is easier to interpret. It shows that judge number 7 differs most from the other judges; he/she has the highest percentage of the loss V_{Within}. Judges 1, 4, 8 and 9 seem to agree quite well, and the other judges take an intermediate position.

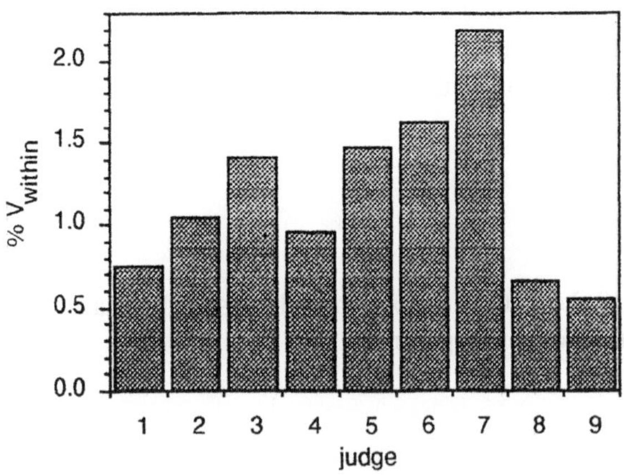

Figure 8 Distribution of the residual V_{Within} variance over the nine judges.

3.6.3 Dimensions

The total amount of variance explained by the two-dimensional consensus-space is 59.5%. Figure 9 shows a scree-graph (compare Figure 5) resulting from a 7-dimensional analysis of the data. From Figure 9 and from Table 11 it can be seen that adding a third dimension to the consensus-space increases the total amount of explained consensus variance to 66.2%.

Table 11 Percentage consensus variance contained in the seven dimensions of the full-dimensional consensus-space of the yoghurt data.

Dimension	$V_{Consensus}$	cum $V_{Consensus}$
1	37.997	37.997
2	19.858	57.855
3	8.390	66.245
4	7.158	73.403
5	5.604	79.007
6	4.566	83.573
7	2.266	85.839
Total	85.838	

A three-dimensional solution explained 68.3% variance. The question of whether or not to include a third dimension depends on the interpretability of a three-dimensional consensus-space.

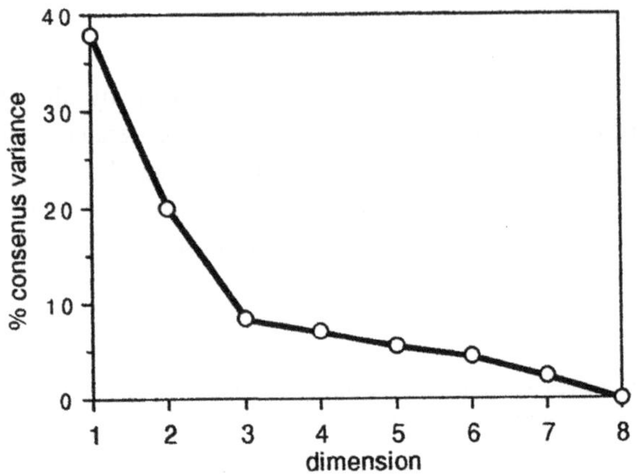

Figure 9 Scree graph of consensus variance per dimension for the full dimensional analysis of the yoghurt data.

3.7 Conclusion

Generalised Procrustes Analysis according to Gower (1975) or to Peay (1988) apply different methods, as shown in Figure 1. The interpretation of the results is the same, noting a difference in sources of variance. The 'Analysis of Variance' tables provided by Generalised Procrustes Analysis software are a useful tool in interpreting the results from the analysis. When the tables report the variances as percentages of the total variance, the entries in the tables correspond with the percentages 'Explained Variance' or percentages 'Variance Accounted For' concept. The distribution of the different variance measured over the products and over the judges provides an indication of the agreement of the judges about particular products and with each other in general. Graphical representation of these tables assists in interpreting the results. The relations between the products can be seen from their position in the consensus-space.

Generalised Procrustes Analysis is a useful technique for the analysis of sensory data, not only for data from Free-Choice profiling experiments, but just as well for data from conventional profiling experiments.

Concluding Remarks Part I: Individual Differences

Summary

In this section the main conclusions from the two chapters in part I are briefly repeated. Suggestions for future research are given, both from the standpoint of sensory research and from the standpoint of Data Theory.

Chapter 2
Assessing Panel Consonance

It is clear that there are large individual differences between the assessors in a sensory panel as there will probably be in any panel. Sensory research may be especially sensitive to idiosyncrasies because of the lack of clear vocabulary for the sense of smell. When a sensory panel is trained to establish a vocabulary it is useful to find out which attributes are "understood," i.e. used consistently by the assessors, and which attributes remain unclear to the panel. The consonance-method proposed in Chapter 2 does just that and can be used to determine the need for further training of either the complete panel or some assessors. The consonance method can be used as a device to monitor the training of a sensory panel and to adapt the vocabulary during the training.

Apart from the individual differences between assessors *within* a panel, there appear differences *between* the panels. Table 1 showed the different ranges of the C-ratios, with a clear anomaly for the steak-panel. These differences may be an effect of different types of training that the assessors received.

In this application of PCA, the finding will often be that there are few attributes completely unidimensional. Considering the following quote from L.L. Thurstone, it is interesting to note that Factor Analysis (or Principal Component Analysis) as applied in this chapter, shows that the domain of sensory and consumer profiling research appears *more* chaotic than it may have looked originally:

> "The purpose of factor analysis.
>
> A factor problem starts with the hope or conviction that a certain domain is not so chaotic as it looks." (Thurstone 1947, p. 55)

Suggestions for Future Research

Other PCA-like techniques can be used for alternative consonance-methods. PCA for categorical data, e.g. Princals (see Gifi 1990), can be used when the data consist of nominal, ordinal or numerical categories, or mixtures of these.

As was already suggested in Chapter 2, the assessor loading-plots could be matched by means of GPA to look for a common structure. The common structure would enable the identification of assessors that are outliers for most attributes, or the discovery of a segmentation of the panel. Especially with a large number of attributes and assessors, when comparison by eye becomes impractical, the matching by GPA would be a useful extension of the method.

The consonance method also provides a means for the comparison of different training methods in order to look for best training and sensory evaluation methods. It may also be useful to compare the effects different laboratories may have.

Another interesting extension to the method would be the comparison of the first eigenvalue with the results from permutation/randomisation analyses or random-data analyses (e.g. Buja and Eyuboglu 1992, Dijksterhuis and Heiser 1995). These methods provide a way to perform a significance test of the first eigenvalue, so it can be seen whether or not a certain attribute is significantly unidimensional, or that there is also a statistically significant second dimension.

Chapter 3
Interpreting Generalised Procrustes Analysis "Analysis of Variance" Tables

This chapter briefly illustrated the differences between the GPA and GPPA methods. The main conclusion is that individual difference models prove useful not only with Free Choice Profiling panels but also with Conventional Profiling panels. Despite the sometimes intensive training the assessors receive, the G(P)PA methods still show that the assessors do not completely agree, and still use attributes differently.

The suggestion to report all Procrustes variances relative to the total variance, hence turning them into Variance Accounted For (VAF) measures, is important, especially since there are two different Procrustes methods available. The way the results of classical Procrustes Analysis were sometimes reported was misleading because the VAF measures were of the final PCA, ignoring the Procrustes-loss. As a result it looked like the GPA results were better in terms of fit than the GPPA results. However, comparison of the two methods in terms of fit is not straightforward (see e.g. Dijksterhuis and Gower 1991/2).

GPA is a very general method which is a useful tool for data-analysts. Especially in the sensory and consumer fields, tailored for the analysis of Free Choice Profiling data, GPA has proven very useful as a research tool (e.g. Oreskovich *et al.* 1991). GPA has become popular in sensory sciences. More recently it is also applied in consumer- and marketing research (e.g. Leemans *et al.* 1992, Steenkamp *et al.* 1994). There is no limit in its use as long as the user realises what she or he is doing. It is advised not just to see GPA as a computer program that gives "an answer."[1] The researcher should always think for her/himself what the method does, and what it is she/he wants to study.

The non-nestedness of GPPA suggests the following scenario for exploratory Procrustes Analysis:

1. perform GPA
2. infer optimal dimensionality by means of the scree-graph
3. perform GPPA in the number of dimensions suggested by 2.

Admittedly, this may be too exploratory for some, but in practice it works satisfactorily. The scenario works because the difference between a

---

[1] The computer-software and manual must be clear about the method that is implemented, otherwise misunderstandings may easily occur. See e.g. the comments of Dijksterhuis *et al.* (1992) on Scriven and Mak (1991) and the responses to it (Scriven 1992, MacFie 1992, Dijksterhuis 1992b).

p-dimensional GPPA solution and the first p dimensions from a GPA solution is usually small. However, note that this is not guaranteed for all data sets.

Graphical Representations

In Chapter 3, the results of the analyses are presented in graphical form, sometimes in addition to their presentation in a table. It is recognised that presentation in both a table *and* a graph is not useful. An exception could be made for the scree-graphs because the "kink" may be easier to see in a graph than in a column of values. The scree in Figure 5 shows the kink at the second dimension. This dimension adds 19.3% variance to the first dimension, which itself explains 45.6%. It can be concluded that the solution is probably best interpreted in one- or two-dimensions, probably not in three-dimensions as is suggested in the chapter. The same critique can be applied to the interpretation of the other scree-graph in the chapter (Figure 9). This graph suggests a two-dimensional solution rather than a three-dimensional one.

The use of the bar-charts in Figure 4 and Figure 8 may be misleading, especially because the caption mentions a "distribution". The charts do not show a "distribution" in the statistical meaning of the term. They were presented only to show the differences between the magnitude of the loss values of the assessors. There is no meaning in the order of the assessors along the horizontal axes of the bar-charts.

Suggestions for Future Research

There are some lines of research in connection with GPA that are interesting, both from an applied Sensory Research viewpoint and from a more theoretical Multivariate Analysis viewpoint.

Statistical Matters

Some researchers expressed their concern about the validity of the results obtained by GPA. This concern can be formulated as in the following question:

> *Is the obtained Group Average meaningful, or is it just a hodgepodge left after intractable mathematical transformations of the data?*

See e.g. the paper by Huitson (1989), and also see the responses by Arnold (1990) and Huitson again (1990). Usually the kind of questions as posed above can be answered by statistics. However, there is hardly any statistics available for GPA methods. Based on certain assumptions, statistical tests may be devised (Davis 1978, Sibson 1978, Langron and Collins 1985), but their usefulness is questionable because the assumptions may be unrealistic. A statistical approach where no distributional assumptions are needed is found in permutation and randomisation tests (see e.g. Edginton 1987). Such tests are recently reported for GPA by King and Arents (1991) and Wakeling *et al.* (1992) for sensory applications. An earlier application of randomisation tests to GPA, as part of a more theoretical PINDIS (Procrustean INdividual DIfference Scaling, Lingoes and Borg 1978) framework was reported by Langeheine (1982).

Another interesting approach is the inclusion in and expansion into a large Analysis of Variance framework, briefly touched upon by Dijksterhuis and Gower (1991/2).

In connection to the question about the meaningfulness of the GPA Group Average, there is a popular misconception about the nature of Procrustes Analysis. This misconception unfortunately pops up every now and then (Huitson 1989, Stone and Sidel 1993).[2] In Chapter 13, *Conclusions*, things are attempted to be straightened out.

---

[2] This misconception was also encountered in several personal communications with researchers in Sensory and Consumer Science.

PART II

MEASUREMENT LEVELS

INTRODUCTION TO PART II

Summary

In this part, non-linear Multivariate models are introduced. Chapter 4 illustrates a variant on classical GPA in which quantitative as well as categorical variables are included in the individual data sets. In Chapter 5, a nonlinear extension to K-sets CCA is introduced. This method is applied to a sensory data set and can be seen as an alternative method to GPA. In Chapter 6, this same method is again introduced, though different, and used for the analysis of another data set from sensory research.

Chapter 4
Multivariate Analysis Of Coffee Images: A Study in the Simultaneous Display of Multivariate Quantitative and Qualitative Variables for Several Assessors[1]

The chapter discusses a small image study in which seven assessors judge nine brands of coffee in terms of six quantitative variables and five categorical variables. The coffees were not tasted — only the package of coffee was presented to the assessors. The data were collected with the intent to use them as an illustration of the methods of data analysis used in this chapter. Generalised Procrustes Analysis and Generalised Biplots are combined to simultaneously display information on the brands and on both quantitative and categorical variables.

The analyses performed in this chapter can be summarised as:

$$\mathbf{X}_k \xrightarrow{\text{distance}} \mathbf{D}_k \xrightarrow{\text{map}} \mathbf{Y}_k \xrightarrow{\text{match}} \mathcal{M}(\mathbf{Y}_k) = \mathbf{Y} \tag{1}$$

where the raw data $\mathbf{X}_k$ are converted into distances in $\mathbf{D}_k$ by a distance generating function g (see formula 1 in Chapter 4), $g(\mathbf{X}_k)=\mathbf{D}_k$. Coordinates of N points in $\mathbf{Y}_k$ can be found by a mapping from $\mathbf{D}_k$ to $\mathbf{Y}_k$, e.g. by means of some form of Multidimensional Scaling. The N points in the rows of $\mathbf{Y}_k$ reproduce the distances in $\mathbf{D}_k$. The K configurations $\mathbf{Y}_k$ have arbitrary orientations, and can be subsequently matched by means of a matching procedure M such as GPA. After a GPA, the configurations $M(\mathbf{Y}_k)=\rho_k\mathbf{Y}_k\mathbf{H}_k$ have commensurable orientations and can be averaged to form a group average configuration $\mathbf{Y}$. When the last step, the averaging, introduces a large amount of loss, the group average must be interpreted with caution, because it is not a good representation of most individual configurations.

The scheme presented in (1) is very flexible. It allows for a large number of different analyses of K data sets, provided that all the sets contain the same N objects in their rows. Some possibilities:

[1] The paper is printed in Quality and Quantity (Gower and Dijksterhuis 1994). An older version of the paper appeared as a research report of the department of Datatheory (Gower and Dijksterhuis 1992) and was presented by the first author at the 3rd Agro Industrie & Methodes Statistiques Conference held in Montpellier in November/December 1992.
The data were especially collected for this paper by OP&P for which Jeannette Timmermans and Pieter Punter are thanked. Jacques Commandeur (department of Datatheory, University of Leiden) kindly did the GPA for data matrices with missing rows.

- making distances using g: Euclidean-, Chi-Square-, Minkowski-p-, Extended Matching Coefficient, etc.
- mapping: PCA, PCO, MCA, MDS, etc.
- matching using M: GPA, GCA, etc.

In this chapter, there is a complication in that the data contain both qualitative and quantitative variables. These two kinds of variables are treated differently. For the quantitative variables, g is taken to produce Euclidean distances, and for the qualitative variables, g is taken to be the Extended Matching Coefficient (EMC, formula 3 in Chapter 4). The Euclidean distances and the Extended Matching Coefficients are combined into one matrix (formula 8, Chapter 4). This matrix is decomposed using PCO/classical scaling, which is the mapping method used. The matching is done by means of GPA.

In fact, three different analyses are performed and their results presented:

1. Categorical variables only (subscript c in (2), see §4.4.1):

$$\mathbf{X}_{k_c} \xrightarrow{\text{EMC}} \mathbf{D}_{k_c} \xrightarrow{\text{PCO}} \mathbf{Y}_{k_c} \xrightarrow{\text{GPA}} \mathcal{M}\left(\mathbf{Y}_{k_c}\right) = \mathbf{Y}_c \tag{2}$$

EMC (formula 3 in Chapter 4), PCO (Chapter 4, Figure 1), GPA (Chapter 4a, Table 4, Figure 2)

2. Both categorical and quantitative variables (respective subscripts c and q, subscripts omitted after their combination, i.e., after the brace in (3); see §4.4.2):

$$\left.\begin{array}{l}\mathbf{X}_{k_c} \xrightarrow{\text{EMC}} \mathbf{D}_{k_c} \\ \mathbf{X}_{k_q} \xrightarrow{\text{Euclidean}} \mathbf{D}_{k_q}\end{array}\right\} \xrightarrow{\text{PCO}} \mathbf{Y}_k \xrightarrow{\text{GPA}} \mathcal{M}\left(\mathbf{Y}_k\right) = \mathbf{Y} \tag{3}$$

EMC, GPA (Chapter 4, Table 4b, Figure 3).

3. As point 1 above but with a joint GPA (§4.4.3, Table 4c, Figure 4).

The GPA's with 1 and 2 above are the usual GPA's where the configurations of the N objects in the rows are matched. The so-called *joint* GPA is performed on the configurations of the N points *and* the 25 category-points. So the GPA is carried out on K matrices with N+25 rows. To handle missing categories — some of the matrices have missing rows — the GPA developed by Commandeur (1991) was used.

Chapter 5
Nonlinear Canonical Correlation Analysis of Multiway Data[2]

The method of nonlinear canonical correlation analysis for K sets of variables, or *Overals*, is introduced. It is both a generalisation of homogeneity analysis (Multiple Correspondence Analysis) as well as of linear canonical correlation analysis. The method is presented from both viewpoints. The formulations via MCA (Chapter 5, formula 5) and via Canonical Correlation Analysis (Chapter 5, formula 7) are shown to be equivalent. It turns out that the inclusion of optimal scaling into K-sets CCA is equivalent to K-sets MCA with additivity constraints and optimal scaling. The mathematics was presented in more detail by van der Burg and de Leeuw (1987).

In this chapter, the method, abbreviated GCA (Generalised Canonical Analysis), is applied as an alternative to GPA. The advantage of GCA is that nonlinearities can be modeled, i.e. nominal, ordinal or both kinds of variables can be analysed in addition to numerical variables. GCA is applied to a data set from sensory research. The data originally consisted of numerical data, but are recoded into just three categories to illustrate the use of ordinal GCA. Three categories is the minimum number of categories to which an ordinal analysis can be applied. The three categories are believed to code the approximate positions of the anchors of the line-scales used in the experiment and of a middle neutral category. There were two anchors, a left anchor ("low" or "bad," depending on the attribute) and a right anchor ("high" or "nice," depending on the attribute).

Just as in Procrustes Analyses, the data consist of K individual sets, where each set represents an assessor. The K sets are matrices with the products in the rows and the attributes in the columns (see also §1.5.3 and Chapter 1, Figure 6).

The data are from a sensory experiment in which a number of smoked sausages are judged by a sensory panel.

---

[2] The paper is written together with Eeke van der Burg (van der Burg and Dijksterhuis 1989) and was presented by the first author at the Multiway 1988 conference in Rome, Italy (see Coppi and Bolasco 1989). The sensory data were analysed before by the second author in his master's at the University of Utrecht (Dijksterhuis 1987).

Chapter 6
Nonlinear Generalised Canonical Analysis: Introduction and Application from Sensory Research[3]

In this chapter, the same technique as in Chapter 5 is introduced. Until then, Generalised Canonical Analysis was described more or less scattered throughout a number of rather technical papers (Carroll 1968, van der Burg *et al.* 1988, van der Burg 1988, Gifi 1990) and it was felt that a concise introduction of the method would be useful. The equivalence of linear GCA with Carroll's (1968) method is shown.

Optimal scaling is introduced as an extension of GCA to include nonlinear transformations. This extension gives the possibility to analyse nominal or ordinal data together with data on an interval level. This presentation shows some overlap with that in the previous chapter.

In this chapter there is mention of a limitation of the Overals software program used, concerning the limited number of categories the program accepts. This is probably only a problem with the old PC versions of the program.

GCA is applied to data from sensory research. Forty different vegetable soups were judged by 19 assessors using 5 attributes. Two different GCA analyses are performed:

- a numerical ten-category solution
- an ordinal three-category solution

The results of these two analyses are compared by matching the two configurations using a Procrustes rotation. The high agreement of the configurations is the reason that only the configurations from the ordinal analyses are presented.

---

[3] The paper is written together with Eeke van der Burg and presented by the first author at the SMABS 1992 conference in Nijmegen (van der Burg and Dijksterhuis 1993b). The paper appeared in the book of the SMABS-conference (Oud and van Blokland-Vogelesang 1993). The data were kindly made available by Dr. B. Cramwinckel (RIKILT-DLO, Wageningen, the Netherlands).

CHAPTER 4

Multivariate Analysis of Coffee Images: A Study in the Simultaneous Display of Multivariate Quantitative and Qualitative Variables for Several Assessors

---

By John Gower and Garmt Dijksterhuis. Originally published in 1994 in *Quality and Quantity*, 28, 165–184. (Reprinted with kind permission from Kluwer Academic Publishers.)

CHAPTER 4

Multivariate Analysis of Coffee Images

A Study in the Simultaneous Display of Multivariate Quantitative and Qualitative Variables for Several Assessors

4.1 Introduction

In consumer research, a panel of assessors is often asked to give judgements on the characteristics of some product under consideration. Judgements may be of diverse kinds and the scales used may include both quantitative and qualitative measurements. Quantitative scales are rarely measured directly, but the members of the panel will be asked to use, say, a ten-point scale or to indicate the strength of their response to some question by marking an appropriate distance along a line. Qualitative responses are selected from a list of attributes (e.g. red, green, yellow, blue for colour). Often attributes are ordered (e.g. low, middle and high income) but in the following we take no cognisance of this information. Occasionally each individual member of the panel chooses his or her own attributes (e.g. in Free Choice Profiling, FCP, Williams and Langron 1984). Rather than write about attributes and characteristics, we use the standard statistical methodology, referring to qualitative variables as categorical variables and to their attributes as category-levels. With FCP not only do the variables differ between individuals, but also the number of variables differs, thus making it difficult to compare the responses of different individuals.

Generalised Procrustes Analysis (GPA, Gower 1975) offers a way of comparing individuals who judge products using different variables. Its basic assumption is that a distance can be defined so that distances between two products, as judged by two individuals, are comparable, even though they may be based on different variables. When the variables are nominally the same, there is no guarantee that two individuals perceive them in the same way, thus supporting the use of GPA in a wider context than just with FCP-like data (e.g. Chapter 3, Dijksterhuis and Punter 1990). This is the justification for using GPA in the following, where the variables used by all assessors are nominally the same. The distance assumption allows a map to be made of the products for each individual. Sets of maps may be matched, in an obvious least-squares sense, and compared as is described in Section 3.

In the use of GPA described here, the distance between pairs of products is defined (see Section 4.3) as a function of the values taken by the quantitative and/or categorical variables used. The original description of GPA did not discuss how information on the original variables could be included in the maps, but soon (Arnold and Williams 1985) quantitative variables were expressed as vectors through a common origin in what amounted to an application of classical biplot methodology (Gabriel 1971). Methods such as multiple correspondence analysis (MCA) show information on categorical variables but only when all the variables have qualitative form and only for one individual. Recent advances in biplot theory (Gower and Harding 1988, Gower 1992a, 1992b) unify the treatment of quantitative and categorical variables and offer other generalisations; for example, MCA is a special case. The methodology used in this paper brings together GPA and a useful special case of Generalised Biplots (GBP, Gower 1992b).

4.2 Data

Seven assessors were asked to judge nine brands of coffee on five categorical and six quantitative variables. The assessors, two men and five women, were presented with a package of each coffee and were asked several questions. The coffees were not tasted, but the assessors were asked to respond according to their conceptions of the properties of the coffee. Thus their responses would be based on their pre-knowledge of the coffees as elicited by the packaging itself. Table 1 briefly describes the nine coffees together with their prices; Table 2 presents the questions.

Table 1 The nine kinds of coffee used in the experiment with their price in Dutch Guilders per 250 grams. The symbols represent the corresponding coffees in the figures.

Symbol	Coffee	Description	Price
△	Red brand	Ordinary coffee	2.54
★	Golden brand	Luxury coffee	3.64
+	Moccona	Instant coffee	12.59
×	Nescafé	Instant coffee	14.25
□	B-Brand	Cheap coffee	1.25
○	Hag	Decaffeinated coffee	3.13
◇	Max Havelaar	Third world coffee	3.58
■	Espresso	Espresso coffee	3.69
●	Chocolate flavour	Coffee with chocolate flavour	11.75

To answer question 6, the assessors were asked to give the price in Dutch Cents (100 Cents = One Guilder) they would be willing to pay for 250g of the coffee. Questions 7 to 11 were scored on line-scales, giving scores ranging from 0 to 100. Each subject was asked the same questions but, for the reason given in Section 4.1, it was thought appropriate to treat the data as if they were of FCP form.

Data of these kinds are collected in the hope of answering questions about the homogeneity, or otherwise, of the patterns of response across individuals. With homogeneous patterns, supplementary problems are to suggest which variables are important and which unimportant in determining the responses, and to examine the extent of departures from an average response. With heterogeneous patterns, it is of interest to ask if there is evidence that the individuals fall into two or more homogeneous groups. Because we had so few assessors, there was insufficient information to answer most questions of these kinds and this exposition must be regarded as a pilot-study to validate the methods proposed.

Table 2 The questions asked of the assessors. The abbreviations of the category-level names can be found in the figures. The underlined parts of the lower table can be found in the corresponding figures.

Questions on categorical variables		Category-levels	
1	How often do you Drink this coffee?		
		Dn	never
		Ds	sometimes
		Dr	regularly
		Do	often
		Da	always
2	What is the most suitable Moment for this coffee?		
		Mb	with breakfast
		Mm	the morning
		Ml	with lunch
		Ma	the afternoon
		Md	after dinner
		Me	the evening
3	What is the most suitable Occasion for this coffee?		
		Oh	at home (each day)
		Ow	at work
		Ov	during vacations
		Or	in a restaurant/café
		Op	at week-ends/public holidays
		Od	after dinner
4	Which Income-group buys this coffee?		
		Il	low incomes
		Im	middle incomes
		Ih	high incomes
5	Would you Buy this coffee?		
		Bn	never
		Bs	sometimes
		Br	regularly
		Bo	often
		Ba	always

Questions on Quantitative Variables		
6	What <u>price</u> are you willing to pay for 250g of this coffee?	
7	Amount of <u>odour</u>?	weak - strong
8	Amount of <u>taste</u> or aroma?	weak - strong
9	Full-flavouredness/<u>rac</u>iness?	weak - strong
10	<u>Bitter</u>ness?	weak - strong
11	<u>Quality</u>?	bad - good

4.3 Methodology

We may imagine data-matrices $\mathbf{X}_k$ (k = 1,2,...,K) to be available for K individuals. A typical element of $\mathbf{X}_k$ will be written x_{ijk}, where i = 1,2,...,N refers to N objects (coffee brands in the example discussed below) which are the same for all K individuals and j = 1,2,...,J_k refers to variables. The matrix $\mathbf{X}_k$ has J_k columns which refer to the variables chosen by the kth individual; in general, these will differ from individual to individual but in our example, J_k is constant. The variables may be quantitative, categorical, or a mixture of both types.

Distances d_{rsk} between all pairs r,s of objects are defined (see §4.3.3) between rows r and s of $\mathbf{X}_k$, where r,s = 1,2,...,N. These may be collected into a symmetric matrix $\mathbf{D}_k = \{d_{rsk}\}$ with zero diagonal. From $\mathbf{D}_k$ a configuration with coordinates given by the rows of a matrix $\mathbf{Y}_k$ may be obtained by any desired form of metric or non-metric scaling. In our example we use principal coordinate analysis PCO/classical scaling which, with the distances used here, always gives real Cartesian coordinates $\mathbf{Y}_k$ in, at most, (N-1)-dimensions, which generate the distances d_{rsk} exactly.

4.3.1 *Overview of Generalised Procrustes Analysis*

The multidimensional-scaling configurations $\mathbf{Y}_k$ (k = 1,2,...,K) are used as the basis of GPA. That is, orthogonal matrices $\mathbf{H}_k$ and, possibly, scaling factors ρ_k are found which minimise

$$\sum_{h<k}^{K} \|\rho_k \mathbf{Y}_k \mathbf{H}_k - \rho_h \mathbf{Y}_h \mathbf{H}_h\|^2$$

or, what is the same thing, minimise

$$\sum_{k=1}^{K} \|\rho_k \mathbf{Y}_k \mathbf{H}_k - \mathbf{Y}\|^2$$

where $\|\mathbf{A}\|^2 = \mathrm{tr}(\mathbf{A}'\mathbf{A})$ and $\mathbf{Y}$ is the GPA group-average given by

$$\mathbf{Y} = K^{-1} \sum_{k=1}^{K} (\rho_k \mathbf{Y}_k \mathbf{H}_k)$$

Thus, the orthogonal matrices $\mathbf{H}_k$ represent generalised rotations which are chosen to optimise the overall match of the individual configurations to their average. The scaling factors ρ_k allow for the possibility that there are size differences between the configurations; a suitable constraint is chosen to exclude trivial solutions with all $\rho_k = 0$. All this is familiar, but we need to establish the notation and draw attention to the formulation that distinguishes the raw data $\mathbf{X}_k$ from the configurations $\mathbf{Y}_k$. In many applications, the two may be taken to be the same, but we wish to emphasise that this is not a necessary constraint; the formulation adopted here allows distances d_{rsk} to be defined very generally and it is this that allows for the possibility of including categorical variables, as will be described below. When $\mathbf{Y}_k = \mathbf{X}_k$, there is an implicit assumption that Pythagorean distance given by $d^2_{rsk}=\sum_{j=1}^{J_k}(x_{rjk}-x_{sjk})^2$ is used. However distance is defined, some preliminary normalisation of the variables of $\mathbf{X}_k$ may be necessary to ensure commensurability; it is assumed that all necessary pre-scaling of this kind has been done (see e.g. Dijksterhuis and Gower 1991/2). When pre-scaling is applied, there is little justification for including the additional scaling factors, ρ_k.

4.3.2 *Components Analysis, Linear Biplots and GPA*

In a principal components analysis of the data for the kth individual, the biplot methodology for exhibiting variables in the graphical display involves the construction of the component loadings $\mathbf{V}_k$ that satisfy the eigenvalue equation

$$(\mathbf{X}_k'\mathbf{X}_k)\mathbf{V}_k=\mathbf{V}_k\Lambda_k$$

where $\mathbf{X}_k$ is now assumed to be expressed in deviations from the sample-means. Thus $\mathbf{V}_k$ is orthogonal ($\mathbf{V}_k'\mathbf{V}_k = \mathbf{I}$, a unit matrix) and Λ_k is diagonal (the matrix of eigenvalues). The ith object has coordinates given by the ith row of $\mathbf{Y}_k = \mathbf{X}_k\mathbf{V}_k$. The vector plotted for the jth variable is obtained by plotting the point whose coordinates are given by the jth row of $\mathbf{V}_k$, and joining this to the origin. This vector is termed the jth biplot axis. Both the component analysis and the biplot axes are usually plotted in some small number r of dimensions, often $r = 2$, that is obtained by using $\mathbf{V}_{kr}$, the first r columns of $\mathbf{V}_k$. Just as the original axes may have associated scales of measurement, so may the biplot axes. On the biplot axes, one unit of measurement may be taken to be (i) that given by $\mathbf{V}_k$ as described above, which we refer to as the *interpolation scale* and (ii) that given by diag($\mathbf{V}_{k_r}\mathbf{V}'_{k_r}$)$\mathbf{V}_{k_r}$, which we refer to as the *prediction scale*. The scales given by (i) and (ii) define just one unit of measurement, and just as when plotting with conventional axes, a series of *markers* may be associated with the biplot axes indicating one unit, two units, three units and so on. To *interpolate* a new sample or product, take the vector-sum of the markers on the interpolation scales representing the values required for the variables. To *predict* the values associated with a sample-point in the ordination, drop perpendiculars from the point and read off the values against the prediction scale markers. In exact representations, the two scales are arranged so that interpolating a set of predicted values recovers the original sample-point and the predicted values to be associated with an interpolated point are the same as the values used for the interpolation; the two scales are consistent. This consistency property is lost in approximations (see Gower 1992a for a more detailed explanation).

Now, suppose biplot axes have been computed as described. Then we can regard the coordinate points on the biplot axes given by the rows of $\mathbf{V}_k$ as being rigidly embedded in the components space, together with the object-points $\mathbf{Y}_k = \rho_k\mathbf{X}_k\mathbf{V}_k$. When GPA rotates $\mathbf{Y}_k$ through $\mathbf{H}_k$, then $\mathbf{V}_k$ is rotated to $\mathbf{V}_k\mathbf{H}_k$ giving biplot axes embedded in the configurations generated by the GPA. Corresponding to the group-average $\mathbf{Y}$ of objects is the group average $\mathbf{Z}=K^{-1}\sum_{k=1}^{K}\mathbf{V}_k\mathbf{H}_k$ which gives the biplot axes to be associated with the group-average configuration. This, however, is valid only when the same variables pertain to all the individuals. With FCP, combining incommensurable variables in this way is not valid. Indeed, the dimensions of the matrices $\mathbf{V}_k$ will generally differ from individual to individual. Even when averaging variables is not permissible, two or more variables may be seen to have similar directions, and then one might provisionally regard this as an indication that these variables might pertain to the same, or similar, underlying factors.

4.3.3 Generalised Biplots

The above sketches what is reasonably well-known. The question arises whether or not something similar can be done for forms of multidimensional scaling other than components analysis, and for distances other than Pythagorean. The basic methodology for this extension is given by Gower and Harding (1988), who described non-linear biplots (NLB) that can be used for any form of distance defined on quantitative variables, and by Gower (1992b), who described Generalised Biplots (GBP) which further extends the methodology to include categorical variables. GBP can be very general indeed, but here we confine ourselves to a special case, which itself has a considerable degree of generality. The results required are stated below; derivations and proofs are given in Gower (1992b).

We assume that each variable contributes independently to overall squared-distance. That is:

$$d_{rsk}^2 = \sum_{j=1}^{J_k} g_j(x_{rjk}, x_{sjk}) \qquad (1)$$

where $g_j(.,.)$ is a function that determines the distance generating function between two samples for the jth variable alone. This general form will be required later, but for the present it is assumed that $g_j(.,.)$ is the same function for all variables and hence may be written $g(.,.)$. The form (1) includes the chi-squared distance of MCA, Pythagorean distance and many popular dissimilarity coefficients.

Next, suppose the rows of $\mathbf{Y}_k$ contain the Euclidean coordinates for the kth individual, generated by some form of ordination as described at the beginning of §4.3. Then the coordinates of the marker ξ on the jth biplot axis relative to the axes of the ordination $\mathbf{Y}$ are given by:

$$\mathbf{y}(\xi)=1/2(\mathbf{Y}\mathbf{Y}')^{-1}\mathbf{Y}'(\mathbf{f}+2\mathbf{D}_k\mathbf{1}n^{-1}) \qquad (2)$$

where $\mathbf{f} = \{x_{ij},\xi\}$, is the vector giving the distance of the marker from each of the N objects ($i = 1,2,\ldots,N$).

As ξ varies $\mathbf{y}(\xi)$ will trace out a non-linear trajectory which corresponds to the jth biplot axis. When $g(x_{rjk}, x_{sjk}) = (x_{rjk} - x_{sjk})^2$ for all j, we have Pythagorean distance and if, additionally, PCO is the ordination method that is used, then we have components analysis and classical linear biplots as a special case. In the examples, we assume this special form, so our biplots for quantitative variables are identical with those of the well-known classical linear biplots. However, GBP also allows categorical variables. Distances for categorical variables may be defined in many ways. Here we assume the extended matching coefficient:

$$g(x_{rjk},x_{sjk}) \begin{array}{l} =0 \text{ when } x_{rjk}=x_{sjk} \\ =1 \text{ otherwise} \end{array} \qquad (3)$$

Thus, if the jth variable is, say, a three-level categorical variable of the colours red, green and blue, then two objects contribute zero distance if they are the same colour, and unit distance if they are different colours. With this simple definition, Gower (1992b) shows that the coordinates that represent the jth categorical variable are given by:

$$\mathbf{Z}_j' = \Lambda^{-1}\mathbf{Y}'\mathbf{G}_j(1/N\ \mathbf{C}_j\mathbf{1}\mathbf{1}' - \mathbf{I}) \tag{4}$$

where $\mathbf{G}_j$ is the indicator matrix for the jth variable (i.e. $\mathbf{G}_j(i,l) = 1$ when level l occurs for the ith object, else is zero) and $\mathbf{C}_j = \text{diag}(\mathbf{G}_j'\mathbf{G}_j)$ gives the number of occurrences of the different levels of the jth variable. Here, and in the formulae that follow, there are J_k of these matrices for the kth individual. Thus, when $J_k = J$, a constant, there are JK sets of coordinates when totalled over all individuals.

One may note the similarity between (4) and the formula for category coordinates in MCA:

$$\mathbf{Z}_j' = \Sigma^{-1}\mathbf{Y}'\mathbf{G}_j\mathbf{C}_j' \tag{5}$$

for which

$$d_{rsk}^2 = \frac{1}{J_k^2}\sum_{j=1}^{J_k}\left(\frac{1}{c_{j_r}} = \frac{1}{c_{j_s}}\right) \tag{6}$$

where c_{j_r} (c_{j_s}) gives the number of occurrences of the category-level of the jth variable observed for the rth (sth) sample for the kth individual and $\Sigma^2 = \Lambda$.

A third possibility is to use the GBP methodology with (1) defined by the chi-squared distance (6). This gives:

$$\mathbf{Z}_j' = \frac{1}{J_k^2}\Lambda^{-1}\mathbf{Y}'\mathbf{G}_j\mathbf{C}_j^{-1} \tag{7}$$

which differs from (5) only in replacing the scaling factors Σ^{-1} by $(1/J^2_k)\ \Lambda^{-1}$. However, we agree with the criticism of Greenacre (1988) that chi-squared distance with the implicit upweighting of scarce categories relative to frequent categories given by (6) is not an attractive distance to use with sparse categorical data. For its simplicity, and for other reasons given by Gower (1992b), we prefer the extended matching coefficient and hence used the category coordinates given by (4) in our example. However, the methods based on (4), (5) or (6) have much in common and are included within the general framework. In all cases, the weighted mean of the category-level points of any variable is at the origin/centroid, i.e. $\mathbf{1}'\mathbf{C}_j\mathbf{Z}_j = 0$.

A major advantage of GBP is the way that (2) allows quantitative and categorical variables to be included in the same analysis. All that is required is that in the distance given by (1), $g_j(x_{rjk}, x_{sjk})$ be defined as Pythagorean for quantitative variables and as the extended matching coefficient for categorical variables. These are the definitions we happen to have used; any other combination of distance definitions that might be deemed desirable may be substituted and in extreme cases every term of (1) could be defined differently.

Thus, to combine both categorical and quantitative variables in one analysis, a PCO was carried out on each $\mathbf{X}_k$, defining squared distances by (1) with the first five terms based on the extended matching coefficient and the remaining six on Pythagorean distances. Such a combined analysis reduces to an eigenvalue decomposition of each of the K matrices

$$1/2\ \mathbf{B}_k + \mathbf{Q}_k\mathbf{Q}_k' = \mathbf{Y}_k\mathbf{Y}_k' \text{ with } \mathbf{Y}_k\mathbf{Y}_k' \text{ diagonal} \tag{8}$$

where

$$\mathbf{B}_k = 1/2(\mathbf{I}-\mathbf{N})\mathbf{G}^{(k)}(\mathbf{G}^{(k)})'(\mathbf{I}-\mathbf{N}) \tag{9}$$

with $\mathbf{Q}_k$ being the mean-centred submatrix of $\mathbf{X}_k$ containing the quantitative variables and $\mathbf{N} = \mathbf{I} - \mathbf{1}\ \mathbf{1}'/N$, where $\mathbf{1}$ is a vector of N units and $\mathbf{G}^{(k)}$ is the indicator matrix for all the categorical variables for the kth assessor. That is $\mathbf{G}^{(k)} = (\mathbf{G}_1^{(k)}, \mathbf{G}_2^{(k)}, ..., \mathbf{G}_5^{(k)})$, where $\mathbf{G}_j^{(k)}$ is the indicator matrix for the jth categorical variable as used in (4).

Thus with $\mathbf{Y}_k$ defined by (8), (4) gives the coordinates of the points to plot for the category levels. The corresponding formula for quantitative variables is:

$$\mathbf{Z}_k' = \Lambda^{-1}\mathbf{Y}_k'\mathbf{x}_{jk}\mathbf{x}'_{jk} \tag{10}$$

This will give coordinates for the observed values of the variable x_{jk} (i.e. the jth column of $\mathbf{X}_k$, the data-matrix for the kth individual) and these will be collinear because $\mathbf{Z}_k$, given by (10), has unit rank because it is the product of the two vectors $\Lambda^{-1}\mathbf{Y}_k'\mathbf{x}_{jk}$ and $\mathbf{x}'_{jk}$. Only one point is needed to define each linear biplot axis but (10) gives N collinear points. A single point, representing the marker for one unit of the *j*th variable, has coordinates $\Lambda^{-1}\mathbf{Y}'_k\mathbf{x}_{jk}$; the other markers are equally spaced along the biplot axis. With non-Pythagorean, but Euclidean, distances, the trajectory is non-linear and even the *N* points corresponding to the data-values of a variable given by the general form of (4) and (10) may be insufficient to give good resolution. Then one would have to use (2) to interpolate as many points as were needed for adequate definition (Gower 1992a).

4.4 Analyses

The analyses are presented as follows: first we give GPA analyses for the seven assessors, based on PCO analyses using (i) only the categorical variables and (ii) all the variables; in both cases, information is included on the variables. We conclude with a novel form of GPA which is permissible only when individual assessors use the same category-levels.

4.4.1 Categorical Variables

Table 3(a) gives the percentage variance explained by the first two dimensions of the PCO solutions using only the categorical variables and defining distance by the extended matching coefficient (3). The scores of assessor 3 capture most variance (63.4%) and those of assessor 5 capture least (54.1%) in the first two dimensions. These percentages might appear disappointing, but are typical for work in this field. Figure 1 shows the two-dimensional PCO's for these two assessors. Comparison is difficult because of the arbitrary relative orientations of the two configurations. However, there are clear differences between the relative positions both of the brands and of the category-levels. Two general comments are prompted by Figure 1. First, it can be seen that neither assessor uses all 25 category-levels to describe the coffee; this is typical, but it can make for difficulties in comparing configurations. Second, the three category levels for income, which one would expect to be a clear case of an ordered categorical variables, are shown as at the vertices of a near equilateral triangle by assessor 3 instead of as an approximate linear ordering. This reflects the multidimensional nature of the extended matching coefficient, which allocates equal distance to the difference between low and high income as to the difference between middle and high income. Table 3(a) and also the PCO plots showed no association with gender, so this aspect is not explored further in the following.

Table 3 Percentage explained variance for the first two dimensions of the PCO analyses of the categorical variables (a) and of all variables (b) of the seven assessors. Female f, Male m.

	(a) Categorical Variables				(b) All Variables			
Sex	Assessor	dim 1	dim 2	Sum	Assessor	dim 1	dim 2	Sum
f	1	36.2	21.0	57.2	1	37.2	25.3	62.5
f	2	39.0	23.4	62.4	2	34.3	22.9	57.2
m	3	40.6	22.8	63.4	3	49.0	21.9	70.9
f	4	34.3	23.5	57.8	4	46.4	19.4	65.8
f	5	30.2	23.9	54.1	5	35.7	18.4	54.1
m	6	31.0	23.7	54.7	6	43.4	19.6	63.0
f	7	34.8	22.1	56.9	7	42.5	19.8	62.3

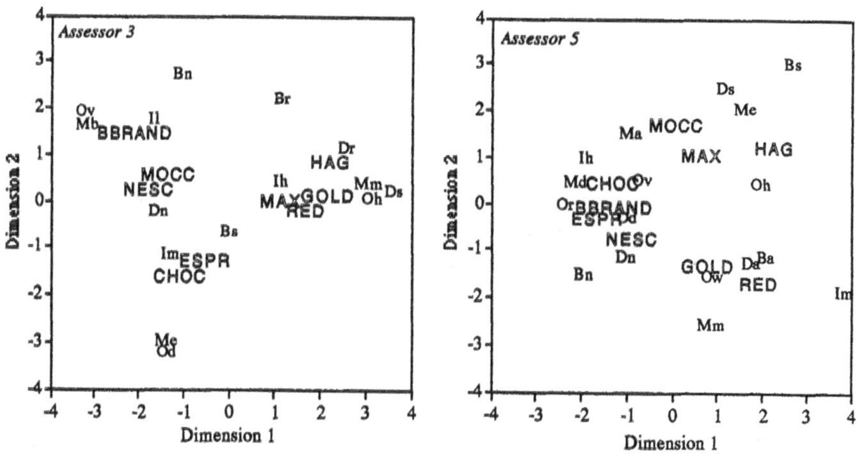

Figure 1 The two-dimensional PCO approximations for assessors 3 and 5 using only categorical variables. See Table 2 for the abbreviations of the category levels.

The role of GPA in minimising the effects of arbitrary orientations is discussed in the following. First, the seven configurations for the brands obtained by PCO are oriented to best fit as described in Section 3; simultaneously, the category-level points were given the same rotations. This could have been done on the two-dimensional approximations, but we have used all eight dimensions required to give an exact representation of the nine brands, so as not to sacrifice information at this stage. To ensure commensurability, the configurations were first scaled to equal sum-of-squares; no isotropic scaling factors were fitted. Figure 2 gives a two-dimensional display for each assessor, where orientation is relative to the principal axes of the GPA group-average and not to its own principal axes; this makes it easier to compare the configurations of the assessors and accounts for the fact that the two-dimensional configurations of assessors 3 and 5 are not rotations of the corresponding configurations of Figure 1. Figure 2 also shows the group-average, which is merely the average of the configurations for the seven assessors. The configurations for assessors 3 and 5 may be compared with those given in Figure 1. As expected, the approximations differ but, apart from orientation, the configurations for assessor 3 have much in common; those for assessor 5 do not agree well, but this is not unexpected as this assessor seems to lie in a different space to the others.

Table 4 Percentage sums-of-squares of the GPA analyses shown in Figures 2, 3 and 4. The variation in the two-dimensions exhibited in the figures is shown separately from the remaining six dimensions. The quantity minimised by GPA is the total residual sum-of-squares. a: Categorical variables only; b: categorical and quantitative variables; c: combined analysis of brands and category-levels.

(a) Categorical Variables Only

	Exhibited 2-Dimensions	Non-Exhibited 6-Dimensions	Total
Group Average	41.47	27.43	68.91
Residual	14.06	17.03	31.09
Total	55.54	44.46	100

(b) Categorical and Quantitative Variables

	Exhibited 2-Dimensions	Non-Exhibited 6-Dimensions	Total
Group Average	49.29	19.60	68.89
Residual	19.74	11.36	31.11
Total	69.04	30.96	100

(c) Combined Analysis of Brands and Category-Levels

	Exhibited 2-Dimensions	Non-Exhibited 6-Dimensions	Total
Group Average	30.34	35.51	65.85
Residual	13.23	20.92	34.15
Total	43.57	56.43	100

The whole of a GPA can be summarised in an analysis of variance and is done in Table 4a where, for convenience, the contributions are expressed as a percentage of the total sum of squares. Everything is in eight dimensions, but it is desirable to separate the variation in the two dimensions exhibited in the figures from the remaining six, and this has been done in Table 4. It can be seen from the total row that the two dimensions exhibited account for rather more than the six dimensions not exhibited and, although dimensionality cannot be equated to degrees of freedom in the normal way, this indicates that Figure 2 is exhibiting about four times as much variation per dimension than is occurring in the unexhibited space. The residual sum-of-squares represents the divergence of the individual assessors from the group-average. The quantity minimised by

GPA is the residual sum-of-squares in the total space, which in this case is 31.09 percent of the total variation, less than half of which is in the exhibited space. The group average, which may be thought of as the "signal," is much better represented in the exhibited space than in the unexhibited space; on a per-dimension basis over four and a half times better. The variation of the group-average could be broken down into the individual contributions of each assessor but, mainly for lack of space, this is not shown. A full account of analysis of variance in the context of GPA and allied methods is given by Dijksterhuis and Gower (1991/2).

Turning again to Figure 2, the same groups of coffees can be identified, although they show much more clearly in Figure 1 than in Figure 2 where they have suffered as a result of the GPA fitting. The configurations in Figure 2 are rotated to best fit the group-average, and when the assessors have heterogeneous responses, the group-average is likely not to show clear differentiation of all the objects by assigning all the objects to positions close to the centre. The detailed structure of the configuration for an assessor is then liable to be diluted in the attempt to match an average that is weak in this assessor's structure. There seems to be a tendency of this kind in the coffee data, but sufficient structure remains to support some tentative remarks. Moccona goes with Nescafé in almost all plots; they are both instant coffees. The cheap B-Brand coffee tends to lie apart from all other coffees, probably reflecting its poor image; it is judged like the instant coffees Mocc and Nesc. In Holland, instant coffee has a poor image. It is used on vacations (Ov) or at breakfast (Mb) by assessor 3. Espresso seems to share properties with Chocolate coffee for assessor 3; it is drunk after dinner (Od), in the evening (Me). The other coffees (Red, Max, Gold, Hag) are sometimes or regularly drunk (Ds, Dr) at home (Oh) in the morning (Mm) by assessor 3. For the same assessor, the categories buy never and buy regularly (Bn, Br) appear at different positions in Figures 1 and 2; apparently these differences are submerged in the comparison with the other assessors by GPA.

When each configuration is compared with the group-average, it can be seen that the assessors do not form a homogeneous group. However, there is some evidence that some category-points apply to some brands rather consistently for all assessors. This can be seen in the Group Average plot, in which the category-points that are used most consistently lie at the outer part of the plot. There seems to be little difference in the occasions 'after dinner', 'at week-ends/public holidays' and 'in a restaurant/café'; the category-points Or, Od, Op lie close together. These occasions apply mainly to the coffees in that part of the figure, i.e. Hag, Choc, and Espr. These coffees are judged to be mainly bought by those with high incomes (Ih) and are never drunk by the assessors themselves (Dn). The instant coffees Mocc and Nesc lie together with D-Brand and are used on vacations (Ov), in the afternoon or with lunch (Ma, Ml), and bought by those with low incomes (Il). These coffees, with Red, are sometimes drunk (Ds) at work (Ow). In addition, Red is bought always or regularly (Ba, Br), and drunk always, often or regularly (Da, Do, Dr). All other coffee brands clutter in the centre of the plot, as do the remaining category-points. Hag (a decaffeinated coffee) is seen to occupy a slightly different position. With only seven assessors, these observations are extremely tentative, especially when one recalls that most assessors used only about two-thirds of the permissible category-levels.

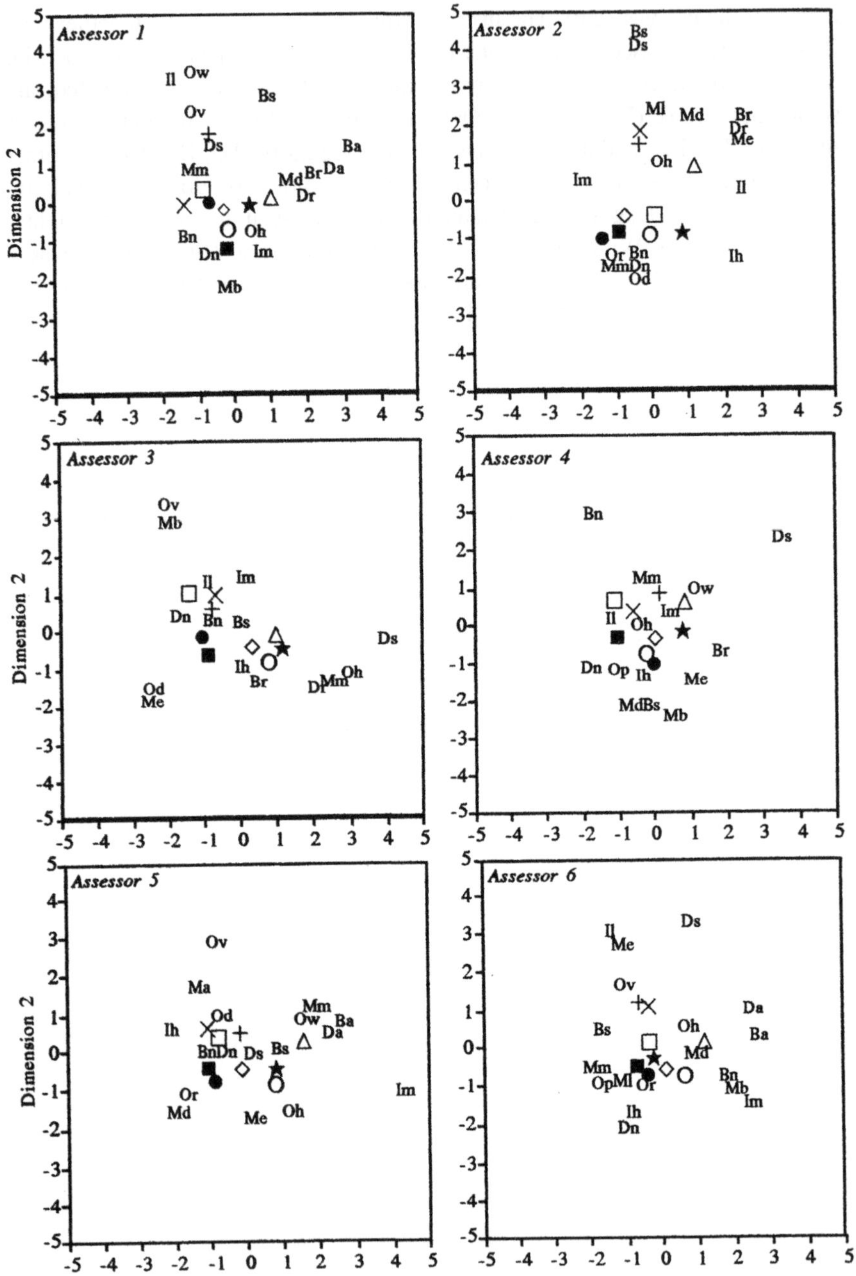
Assessor 1
Assessor 2
Assessor 3
Assessor 4
Assessor 5
Assessor 6
Dimension 2
Dimension 2
Dimension 2

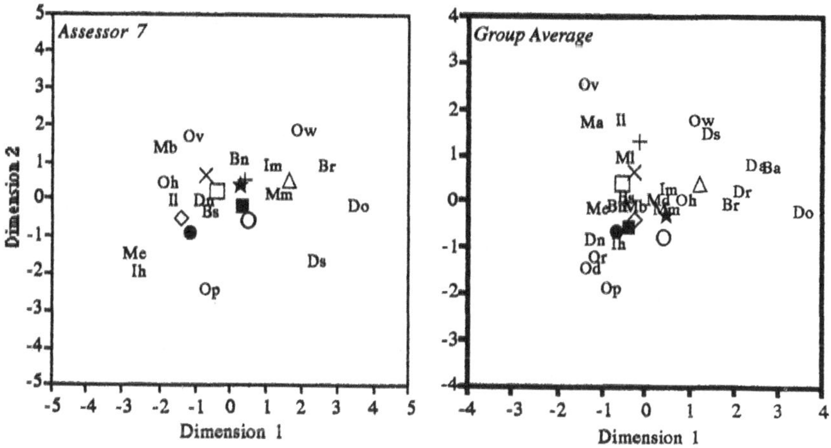

Figure 2 The GPA of each assessor shown relative to the principal axes of the group-average (categorical variables only). Also shown is the group-average itself. The symbols for the coffees are defined in Table 1.

4.4.2 *Quantitative and Categorical Variables Combined*

A similar analysis was done using both quantitative and categorical variables. The percentage explained variance for the first two dimensions has been included in Table 3(b) and the associated analysis of variance is in Table 4(b). It shows a similar level of approximation to that given by the categorical variables alone. Figure 3 is the counterpart of Figure 2 but for a GPA based on all variables. Its main difference lies in the inclusion of quantitative variables which induce linear biplot axes. Because the quantitative variable axes extend beyond the positions for the coffee brands, the brands unfortunately appear superimposed in the centre of the plot; ideally the plots should be enlarged to put Figure 3 on the same scale as that of Figure 2. Apart from this artefact, the positions of the brands in the two figures compare well.

In Figure 3 the quantitative biplot axes are labelled at the high-score end by the corresponding variable with an initial capital letter, and in lower case at the low-score end. These correspond to the assessors' maximal and minimal scores, respectively. Note that the linear biplot axes are not drawn in the Group Average plot. This could have been done in several ways, averaging scale points corresponding to the same raw score, or averaging the scale points that correspond with the minimum or maximum scale points used. The axes could also have been included in the GPA matching process by averaging over assessors the unit-points on their corresponding biplot-axes, but we do not show this in the figures.

When we examine the linear biplot axes for assessor 3 we see that the quantitative variables, quality, racy, taste and bitter, all point in the same direction. These attributes are clearly correlated for this assessor. Odour seems to be a little different and it seems as if price is only partly correlated with these attributes. The right-hand side of the plot is characterised by high scores on the quantitative attributes. For assessor 5 we see that coffee, judged to be of high price, is drunk mostly during vacations (Ov). These characterisations apply to the instant coffees Moccona and Nescafé. For this person, high quality coffees score low on bitter and racy, and high on taste.

The most important category points can be seen in the Group Average plot. In general they seem to be the same as in the group average of Figure 2.

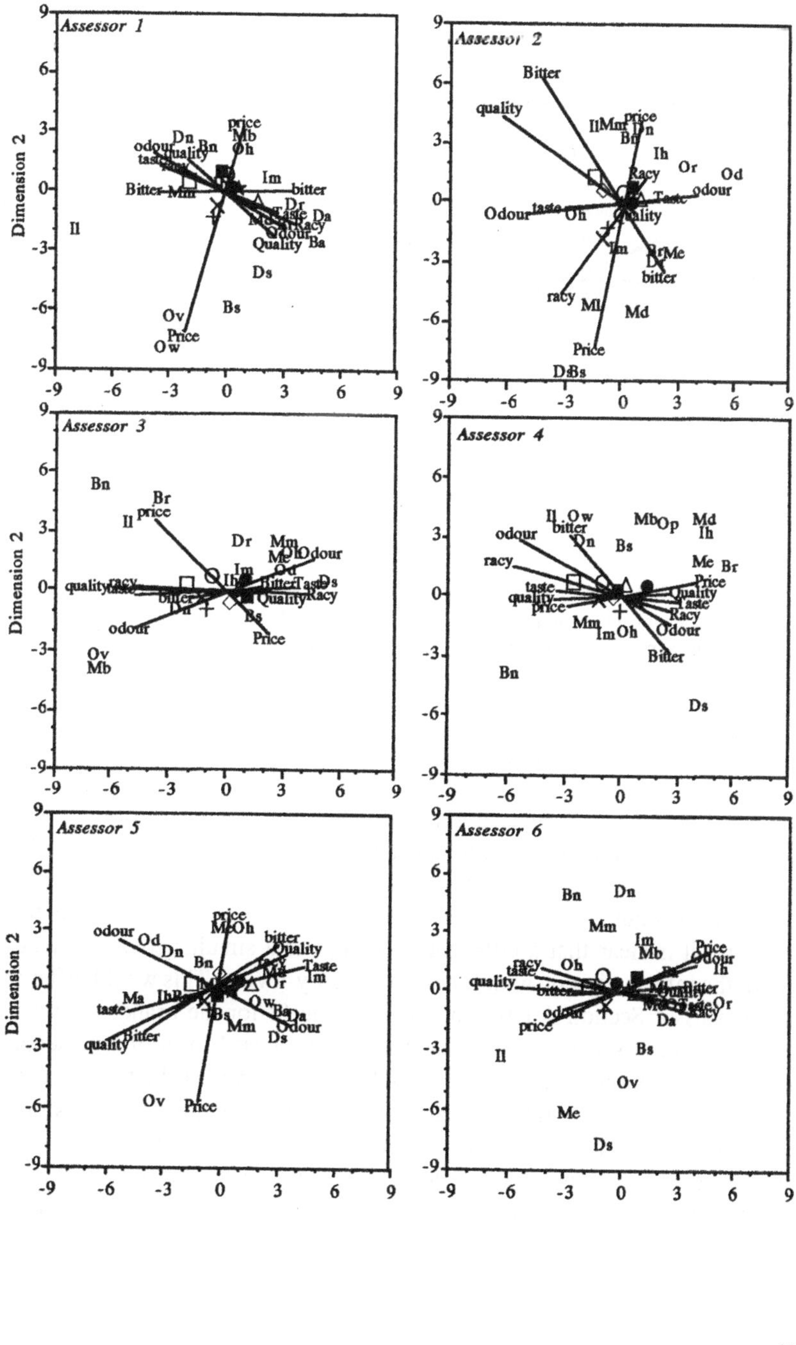
Assessor 1
Assessor 2
Assessor 3
Assessor 4
Assessor 5
Assessor 6
Dimension 2
Dimension 2
Dimension 2

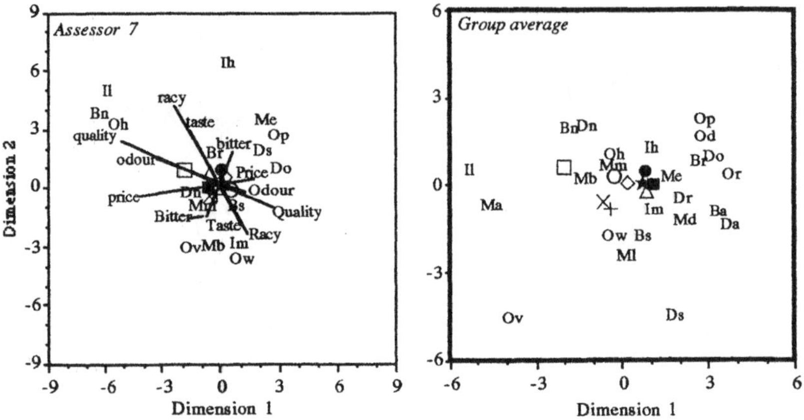

Figure 3 The GPA of each assessor shown relative to the principal axes of the group-average (categorical variables and quantitative variables). Also shown is the group-average itself. The symbols for the coffees are defined in Table 1.

4.4.3 *Joint GPA on Brands and Category Levels*

Returning to the analysis of categorical variables given in Section 4.4.1, we have points for nine brands and a total of 25 category-levels. These are common to all assessors, except that some levels are missing for some assessors. Assuming for the present that all 34 points are available for all assessors, it is clear that a GPA could be done that simultaneously oriented to best fit the information on brands and on category-levels. This would differ from the analysis of Section 4.1 which optimises the fit for the brands, leaving the category-levels to fit in as best they can. The proposed form of representation must give a poorer fit for the brands but it gives a better fit for the category-levels, so it might be regarded as a better compromise to exhibiting both types of information. Figure 4 shows the combined analysis. In doing this analysis, some responses to questions about some categorical variables were missing categories for some assessors. This complicates the computational processes for matching configurations, but Commandeur (1991) has discussed the modifications required and which were used in our analysis.

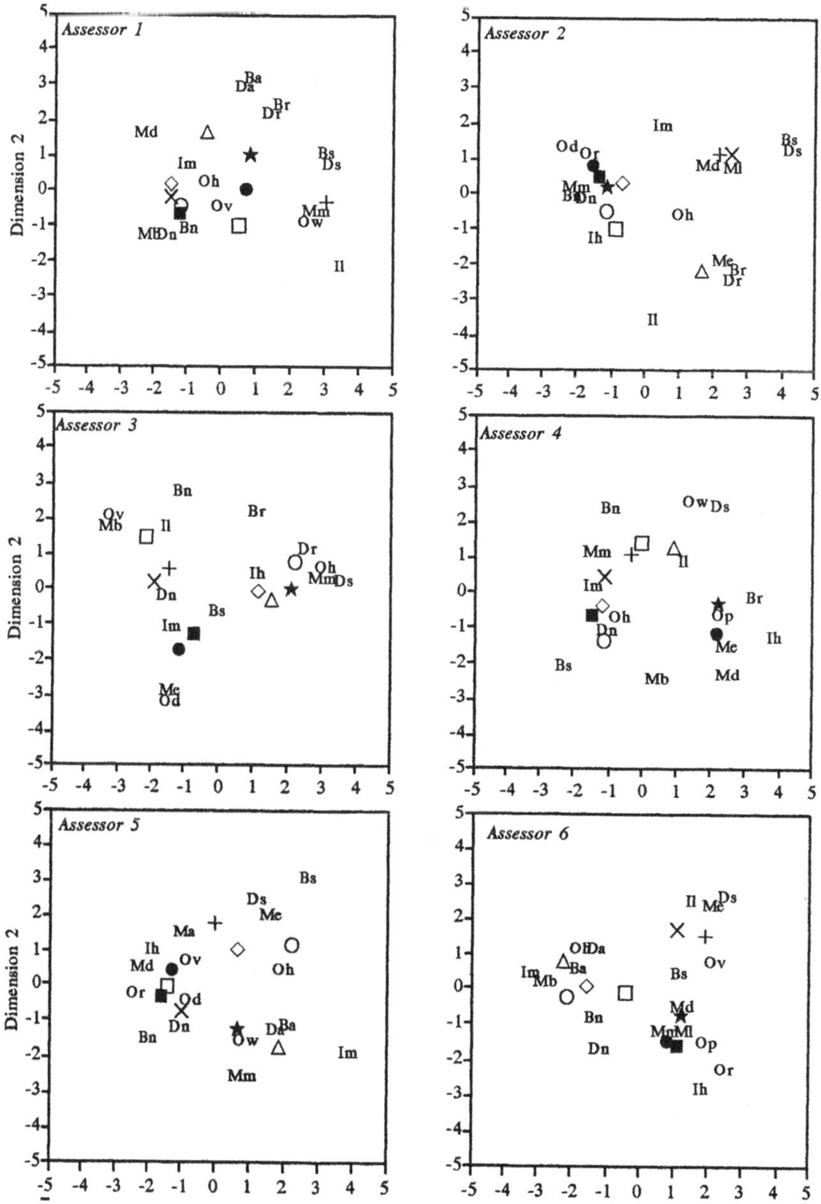
Assessor 1
Assessor 2
Assessor 3
Assessor 4
Assessor 5
Assessor 6
Dimension 2
Ba
Da
Br
Dr
Md
Im
Oh
Ov
Bs
Ds
Mm
Ow
Bn
Il
Od
Or
Ml
Ih
Me
Ma
Op
Mb

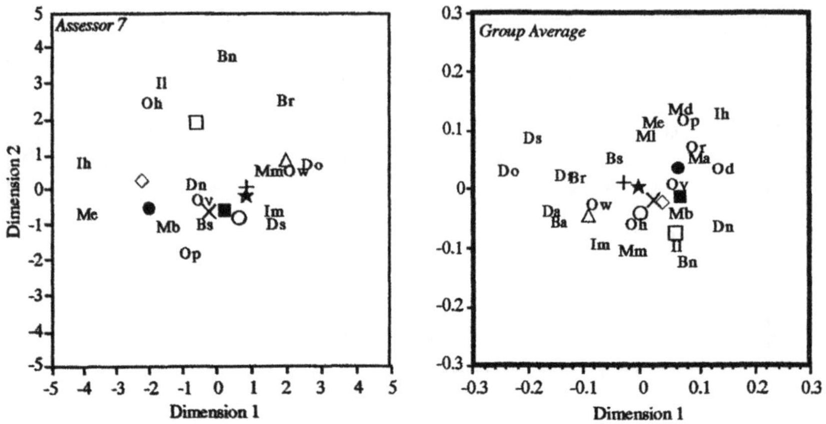

Figure 4 The GPA of each assessor shown relative to the principal axes of the group-average (categorical variables only). Also shown is the group-average itself. In this analysis, optimal fit has been obtained using the points representing the brands and the points representing the category-levels. The symbols for the coffees are defined in Table 1.

By including the category-points in the matching process, it is clear that a better overall spread of both the brands and category-points has been achieved. Comparing the assessors' configurations with the ones in Figure 2 apart from orientation reveals similar configurations for both category and coffee-positions. Table 4c gives the analysis of variance and shows that the exhibited two dimensions give a poorer representation of the coffees. This is the price that has to be paid to accommodate better the information on the category-levels. Nevertheless, on a per dimension basis, the two dimensions are still accounting for much more than the unexhibited dimensions.

Of course, the GPA could also be done solely on the category-level points, thus giving an optimal representation for the variables and leaving the brands to fit as best they can. We could also have done the combined analysis as described in Section 4.2 and include the linear biplots for the quantitative variables. In the latter form of analysis, the linear biplots in the group-average configuration are obtained by averaging like scale points on the axes for the individual assessors. Although both of these analyses might sometimes be useful, we did neither, but that the possibilities exist exemplifies the flexibility of the methodology.

4.5 Conclusion

The main thing that we have done here is to demonstrate the feasibility of this kind of analysis, especially combining information on quantitative and categorical variables. The few assessors that we have used to demonstrate the methodology would always have been inadequate for a serious investigation into coffee images but the apparent heterogeneity of the responses compounds the difficulty in arriving at any firm conclusions. However, the joint analysis described in § 4.4.3 has some attractive features.

In §4.2 it was pointed out that there might be interest in seeing whether there was evidence that the assessors fell into two or more groups. With only seven assessors, it seemed futile to try to answer this question and, in any case, in a GPA a heterogeneous group-average would tend to obscure such differences, if they existed. It would seem better to proceed by accumulating all of the 21 pairwise Procrustes residual sums-of-squares statistics into a 7×7 symmetric distance-matrix and displaying the seven assessors by some form of multidimensional scaling (Gower 1971). Then it can readily be seen if the assessors group or not.

CHAPTER 5

Nonlinear Canonical Correlation Analysis of Multiway Data

5.1 Introduction

The form of nonlinear canonical correlation analysis (CCA) described in this paper is a technique that gives linear combinations of variables within two or more sets, such that these linear compounds are as similar as possible to an unknown orthogonal configuration. At the same time, the variables can be transformed nonlinearly so that nominal or ordinal data can be handled. Transformations are obtained with the help of optimal scaling (cf. Young 1981). The technique is called OVERALS (Gifi 1981, van der Burg, De Leeuw and Verdegaal 1988). It is implemented in a computer program for which the same name is used (Verdegaal 1986). However, from the text it is always clear whether we are referring to the technique or to the program.

There are different ways to get to the formulation of OVERALS. One leads via k-sets homogeneity analysis (or multiple correspondence analysis, MCA) and the other one via k-sets linear CCA. The second route via linear

Originally published in 1989 in: Coppi, R., Bolasco, S. (Eds.) *Multiway Data Analysis*, 245–255. (reprinted with kind permission from Elsevier Science Ltd, the Boulevard, Langford Lane, Kidlington OX5 1GB, UK)

CCA (cf. van der Burg and De Leeuw 1987) leads directly to the definition mentioned above. However, it does not reveal the relationship with correspondence analysis, which the first route does. Therefore we prefer an introduction via MCA. A much more detailed description can be found in van der Burg *et al.* (1988) or van der Burg (1988, Chapter 5). These authors also discuss the algorithm and the solutions for the various parameters.

5.2 *K*-Sets Homogeneity Analysis

Homogeneity analysis is a technique which maximises the homogeneity between a number of variables by assigning values to the various categories (Guttman 1941, De Leeuw 1973, Benzécri 1973, Nishisato 1980, Gifi 1981, Greenacre 1984, Lebart, Morineau and Warwick 1984). Suppose we have an $n \times m$ multivariate data matrix, with rows corresponding to objects and columns to variables. Assume that variable j has k_j different categories and define $\mathbf{G}_j$ ($n \times k_j$) as the indicator matrix corresponding to this variable. Then element (i, l) of $\mathbf{G}_j$ is equal to one if object i belongs to category l, and is zero otherwise. The matrix $\mathbf{Y}_j$ ($k_j \times p$) represents a p-dimensional *quantification* of the categories, and $\mathbf{G}_j\mathbf{Y}_j$ represents a p-dimensional quantification or *transformation* of variable j. The variables are perfectly *homogeneous* if all $\mathbf{G}_j\mathbf{Y}_j$ are equal to a matrix of object scores of $\mathbf{X}$ ($n \times p$). As perfect homogeneity hardly ever occurs, the aim of MCA is to minimise the *loss* of homogeneity (De Leeuw 1973). As an extra condition, the object scores must be standardised (variance one and zero mean for each column) and uncorrelated. Thus MCA is

$$\text{minimize} \sum_{j=1}^{M} \text{SSQ}(\mathbf{X} - \mathbf{G}_j\mathbf{J}_j) \text{ over } \mathbf{X} \text{ and } \mathbf{Y}_j \tag{1}$$
$$\text{subject to the conditions that } \mathbf{X}'\mathbf{X} = n\mathbf{I} \text{ and } \mathbf{u}'\mathbf{X} = 0$$

SSQ means sum of squares, $\mathbf{u}$ is a n-vector of ones, and $\mathbf{I}$ ($p \times p$) is the identity matrix.

MCA treats all variables in an equal way. Suppose, however, that the variables are divided into k sets and let us indicate the indices of set t by $J(t)$. (Note that k and k_j have different meanings). Then one way to express the set structure of the variables is by forming *interactive codings* for all possible combinations of categories for each set. Define $\underline{\mathbf{G}}_t$ ($n \times \underline{k}_t$) as the indicator super-matrix for set t and $\underline{\mathbf{Y}}_t$ ($\underline{k}_t \times p$) as the super-matrix of category quantifications, where $\underline{k}_t = \prod_{j \in J(t)} k_j$. Then k sets homogeneity analysis may be defined as

$$\text{minimize} \sum_{t=1}^{k} \mathrm{SSQ}(\mathbf{X} - \underline{\mathbf{G}}_t \underline{\mathbf{Y}}_t) \text{ over } \mathbf{X} \text{ and } \underline{\mathbf{Y}}_t \tag{2}$$

$$\text{subject to the condition that } \mathbf{X}'\mathbf{X} = n\mathbf{I} \text{ and } u'\mathbf{X} = 0$$

As the number of categories can be huge in each set (four variables with five categories give 5^4 categories for the interactive coding) it is necessary to restrict the number of categories. An obvious way to attain this is by using *additivity constraints*. In the terminology of analysis of variance, this means a restriction to main effects only. This gives

$$\underline{\mathbf{G}}_t \underline{\mathbf{Y}}_t = \sum_{j \in J(t)} \mathbf{G}_j \mathbf{Y}_j \tag{3}$$

Another way to restrict the number of categories, or more correctly the number of category quantifications, is by *rank-one restrictions*. This corresponds to restricting the category quantifications $\underline{\mathbf{Y}}_t$ or $\mathbf{Y}_j$ to lie on a line (i.e. the matrices $\underline{\mathbf{Y}}_t$ or $\mathbf{Y}_j$ have rank one). Rank-one restrictions are supplemented with *optimal scaling* restrictions (i.e. the quantifications have to satisfy the measurement level restrictions). We combine additivity constraints with rank-one restrictions and optimal scaling restrictions, thus we apply these restrictions to $\mathbf{Y}_j$. Then the following holds for one or more variables:

$$\mathbf{Y}_j = \mathbf{z}_j \mathbf{a}'_j \text{ (rank-one) with } \mathbf{z}_j \in C_j \text{ (optimal scaling)} \tag{4}$$

where $\mathbf{a}_j$ is the p-vector of *weights*, $\mathbf{z}_j$ the k_j-vector of *single category quantifications*, and C_j the set of (*standardized*) *admissible quantifications* according to the measurement level of the variable (*nominal, ordinal* or *numerical*). In the nominal case, the rank-one restriction $\mathbf{Y}_j = \mathbf{z}_j \mathbf{a}'_j$ is the only restriction (with $\mathbf{z}_j$ standardised). In the ordinal case, $\mathbf{z}_j$ is a monotone transformation of the original category scores (raw data), and in the numerical case, $\mathbf{z}_j$ is a linear transformation of the original scores. Standardisation of $\mathbf{z}_j$ corresponds to a weighted sum of squares equal to n and a weighted mean equal to zero, the weights being the frequencies corresponding to each category.

OVERALS is defined as k-sets homogeneity analysis with additivity constraints and optimal scaling. It is formulated as follows:

$$\text{minimize} \sum_{t=1}^{k} \text{SSQ}\left(\mathbf{X} - \sum_{j \in J(t)} \mathbf{G}_j \mathbf{Y}_j\right) \text{ over } \mathbf{X} \text{ and } \mathbf{Y}_j$$
$$\text{subject to the condition that } \mathbf{X}'\mathbf{X} = n\mathbf{I} \text{ and } \mathrm{u}'\mathbf{X} = 0$$
$$\text{and for some variables } \mathrm{Y}_j = \mathbf{z}_j \mathbf{a}'_j \text{ with } \mathbf{z}_j \in C_j \tag{5}$$

(cf. Gifi 1981, van der Burg *et al.* 1988). This definition seems not to agree with the one in the first paragraph of the introduction. However, they are the same. We will explain this in the next section.

5.3 *K*-Sets Canonical Correlation Analysis

To understand why OVERALS (5) is a form of generalised (k-sets) CCA let us consider the case in which all variables are optimally scaled. Define $\mathbf{G}_j\mathbf{z}_j$ as $\mathbf{q}_j$ and $\mathbf{Q}_t$ as the matrix with all $\mathbf{q}_j$, $j \in J(t)$, written next to each other, and $\mathbf{A}_t$ as the matrix with all $\mathbf{a}'_j$ written above each other. Then we get

$$\sum_{j \in J(t)} \mathbf{G}_j \mathbf{Y}_j = \sum_{j \in J(t)} \mathbf{G}_j \mathbf{z}_j \mathbf{a}'_j = \sum_{j \in J(t)} \mathbf{q}_j \mathbf{a}'_j = \mathbf{Q}_t \mathbf{A}_t \tag{6}$$

Thus in case all variables are optimally scaled, OVERALS (5) turns into

$$\text{minimize} \sum_{t=1}^{k} \text{SSQ}(\mathbf{X} - Q_t\ \mathbf{A}_t) \text{ over } \mathbf{X} \text{ and } \mathbf{Q}_t$$
$$\text{subject to the conditions that } \mathbf{X}'\mathbf{X} = n\mathbf{I} \text{ and } \mathrm{u}'\mathbf{X} = 0$$
$$\text{and } \mathbf{q}_j \in C_j \text{ with } j \in J(t) \text{ and } t = 1, \ldots, k \tag{7}$$

We use the same notation C_j for the set of admissible quantifications with regard to $\mathbf{z}_j$ as to $\mathbf{q}_j$, as these sets are equivalent (the elements of $\mathbf{q}_j$ are $\mathbf{z}_j$-values, i.e. $\mathbf{q}_j = \mathbf{G}_j\mathbf{z}_j$). The problem formulated in (7) is easily recognised as a form of generalised canonical correlation analysis. When the optimal scaling restrictions consist of only numerical restrictions, the matrices $\mathbf{Q}_t$ are in fact no longer variable. Their columns represent the standardised versions of the columns of the data matrix. Then the formulation of k-sets CCA is equivalent with the one from Carroll (1968). Other criteria in generalising CCA are found in Horst (1961), Kettenring (1971) and van de Geer (1984).

In the literature, the columns of the matrices $\mathbf{Q}_t\mathbf{A}_t$ are called the *canonical variates*. They correspond to the linear compounds of transformed variables mentioned in the introduction. In case there are two sets of variables, the canonical variates are uncorrelated within each set. This does not hold for the analysis of three or more sets as in (7).

The problem formulated in (7) seems more restrictive than (5), as all variables are supposed to be optimally scaled. However, a different look at the p-dimensional quantification of the categories of a variable brings the two definitions together. Every matrix $\mathbf{Y}_j$ can be written as a product of two matrices, e.g. $\mathbf{Z}_j\mathbf{A}_j$. The matrix $\mathbf{Z}_j$ may simply contain the standardised columns of $\mathbf{Y}_j$ and $\mathbf{A}_j$ may contain the corresponding standard deviations (diagonal matrix), but other decompositions are also possible. If we write $\mathbf{z}_{jr}$ for the columns of $\mathbf{Z}_j$ and $\mathbf{a}'_{jr}$ for the rows of $\mathbf{A}_j$, then

$$\mathbf{G}_j\mathbf{Y}_j = \mathbf{G}_j\mathbf{Z}_j\mathbf{A}_j = \sum_{r=1}^{p} \mathbf{G}_j\mathbf{z}_{jr}\mathbf{a}'_{jr} \tag{8}$$

In (8) we are dealing with only one indicator matrix. Thus a p-dimensional quantification $\mathbf{Y}_j$ can be considered as p single quantifications of the same variable. This implies that if we have *p copies* of a variable within its set, and we use optimal scaling (nominal restrictions), we get a p-dimensional quantification. Consequently problem (7) extended with the notion of copies corresponds to (5).

The definition of OVERALS given in the introduction corresponds literally to (7). With the help of copies this definition also corresponds to (5). Thus k-sets CCA with optimal scaling and k-sets homogeneity analysis with additivity constraints and optimal scaling are similar. They are both synonymous to OVERALS. It is a matter of preference which definition one wants to use. A more detailed description of OVERALS interpreted as k-sets CCA is given by van der Burg and de Leeuw (1987).

The algorithm of the OVERALS computer program is of the alternating least squares (ALS) type. This means that parameters are solved for alternatingly, keeping the other parameters at a constant level. This is an iterative process. Each step consists of solving a least squares problem (van der Burg *et al.* 1988).

For the OVERALS computer program, usually data matrices of the form objects × variables (divided into sets) form the input. However, also multiway data with three levels can be analysed by OVERALS. Suppose the structure of the data is: objects × variables × occasions. If a researcher is interested in the similarity between the occasions, he may consider using OVERALS. In multiway data we always deal with similar variables and objects for all occasions. For the OVERALS technique this is not necessary with regard to variables, i.e. objects may be measured for sets (occasions) of different variables.

5.4 An Application of Overals to Multiway Data

An application of OVERALS to a multiway table is taken from the field of sensory research. In sensory research, the assessment of products by judges is studied. Typical sensory data are three mode data, the three modes being products, judges and aspects. We are interested in the (perceived) quality of certain aspects of the products. In this particular example the products are sixteen different brands of smoked sausages, judged by ten untrained persons (see also van Buuren 1987, Dijksterhuis 1987). Often in sensory research, panels are trained to attain consensus about the meaning and the use of the aspects (from now on called variables). But even then one has to assume that all individual judges interpret each variable in a similar way, e.g., when tea is judged on bitterness one can safely make this assumption but in describing perfumes a variable like "feminine" is likely to have a different meaning to each judge. In the sausage example we are dealing with untrained judges and therefore we will not assume a consensus about the variables. This is the reason why we are interested in an OVERALS analysis of these data, as OVERALS does not assume similar variables in each set.

The variables used in this experiment describe certain attributes of the sausages. The attributes used for this study are: "appearance" (bad to nice), "taste" (bad to nice), "odour" (bad to nice), "price" (cheap to expensive) and "nasty taste" (little to much). The judges were presented with a piece of sausage and rated the attributes on a line scale. The original data ranged between 0 and 50 and were divided into three categories (0-20, 21-29, 30-50). In the terminology of judges the lowest category reflects the left anchor of the line scale (e.g. "bad"), the highest category reflects the right anchor (e.g. "nice"), and the middle category corresponds to a neutral position. The obtained categories were assumed to have ordinal properties.

Since we are interested in the similarity between the judges, we use OVERALS to construct a common object space for the 10 judges. Each judge defines a set of 5 variables. We thus obtain 10 sets. In total there are 49 variables (one variable was excluded from the analysis as it did not vary after categorisation). We want to obtain a two-dimensional solution for the sake of easy graphical display.

The OVERALS analysis provided an eigenvalue of 0.70 for the first dimension and of 0.66 for the second dimension. The eigenvalues are a measure of the fit per dimension, which can be maximally 1.0. The OVERALS-loss of (7) is a function of the eigenvalues. Details can be found in van der Burg *et al.* (1988) or van der Burg (1988, Chapter 5). From the eigenvalues we conclude that the sets are quite different. This means that the judges use the variables in rather different ways.

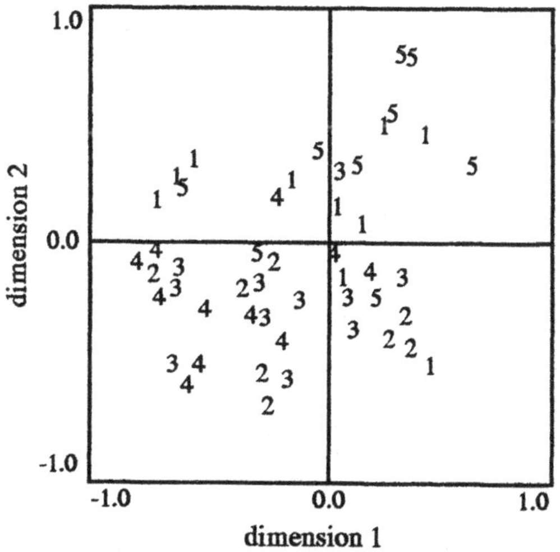

Figure 1 Component loadings for all 49 variables (1=appearance, 2=taste, 3=odour, 4=price, 5=nasty taste).

In Figure 1 the component loadings for all 49 variables are presented in the two-dimensional canonical space (space of object scores). The component loadings correspond to the correlations of the transformed variables with the object scores. The variables are labelled 1 to 5 for each judge. It can be seen from Figure 1 that there does exist some agreement about the use of the variables between the judges. But it is also clear that individual differences do exist, since most variables with equal numbers are spread over two quadrants. To take a closer look at the use of the variables by the individual judges, we present the component loadings per set (i.e. per individual judge) for four sets in Figure 2. (Only four sets are presented here.)

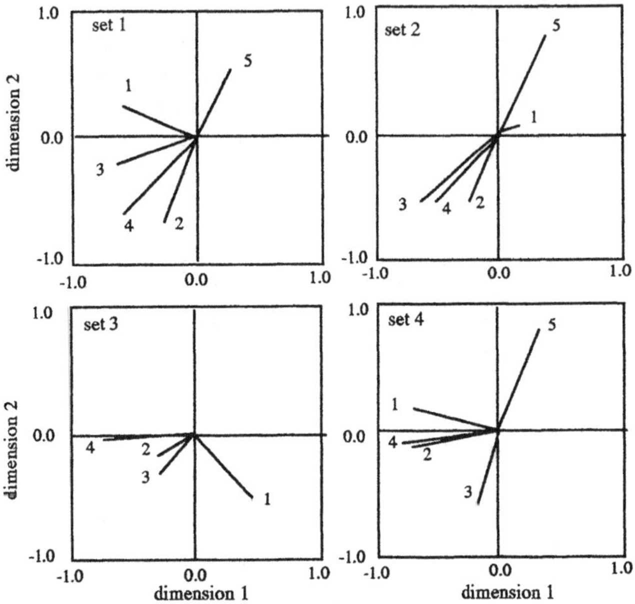

Figure 2 Component loadings for four sets (1=appearance, 2=taste, 3=odour, 4=price, 5=nasty taste).

Figure 2 shows that strong agreement exists between the sets 1, 2 and 4 with respect to variable 5 ('nasty taste') (for set 3 this was the excluded variable). For sets 1 and 2 (and 5, 6 and 8 not shown here) the variables 2 and 5 point to opposite directions. Obviously, sausages with nice taste do possess little nasty taste. However, for some judges this is not so clear, judge 4 (and 7, 9 and 10 not shown here) uses this variable somewhat differently. Other things that can be seen from Figure 2 are e.g.: For judge 2 the variables 'odour', 'taste' and 'price' are correlated, while 'appearance' seems to exert very little influence. 'Nasty taste' is important and lies opposite to the other variables. For judge 4, 'price' and 'taste' are almost similar, 'appearance' lies in the 'expensive' and 'nice taste' direction. Again 'nasty taste' is directed away from the other variables. 'Odour' seems to have some particular meaning for this judge. It points to a direction different from 'taste', 'price' and 'appearance', and opposite to 'nasty taste'.

From these plots we learn about the use of the variables by the judges, as this is related to the role the variables played in the analysis. We can identify judges using variables in a deviant way.

Another way to look at the variables is by inspecting the single category quantifications. In Figure 3 the single category quantifications are given for all variables and all sets. We assumed all variables to be of ordinal measurement level and we can check the effect of this assumption with the help of the single category quantifications (Figure 3). When the quantifications are like the one of variable 3 for judge 8, numerical restrictions would have given the same analysis results. When, on the other hand, most transformations behave like the ones of variables 1, 2 and 3 of judge 7, the ordinal assumption is rather restrictive. Here two categories are scaled at the same value, in this case the order of the category quantifications may change when a nominal measurement level was assumed. However we prefer an ordinal interpretation of the data. Furthermore we see that for all judges (except no. 8) the highest category of 'nasty taste' is scaled at a rather high value. As all quantifications are standardised, this means that the highest score for 'nasty taste' was less often used than the highest scores for the other variables. From Figure 3 we also see that judge 4 only used extreme and no neutral categories of the variables.

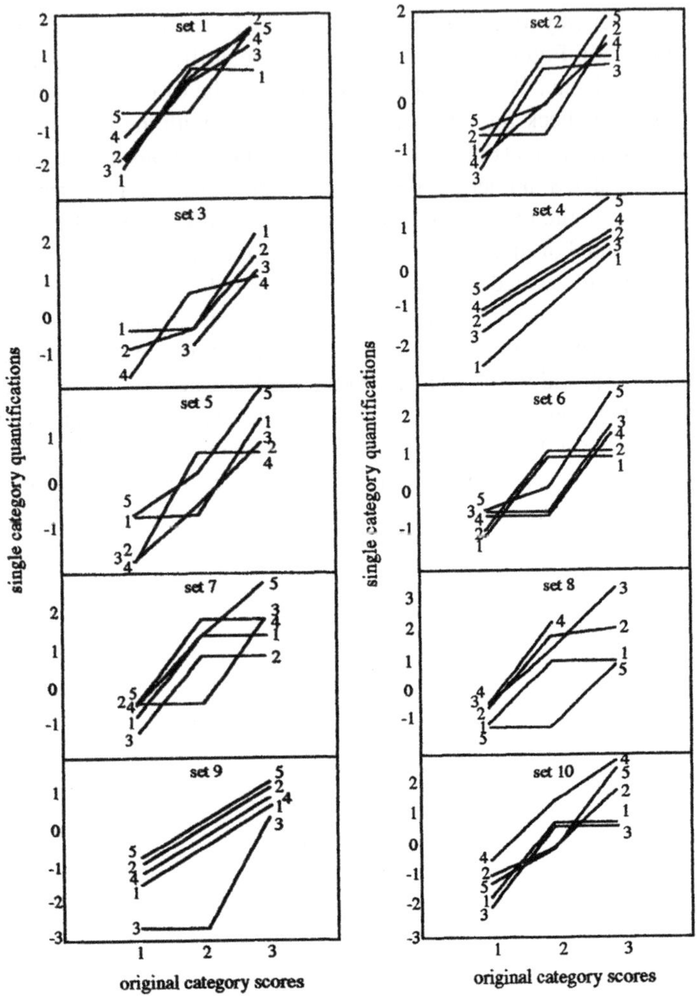

Figure 3 Single category quantifications per set (1=appearance, 2=taste, 3=odour, 4=price, 5=nasty taste).

In Figure 4 the sausages are plotted in the canonical space by means of the object scores. In the plot, a clear distinction exists between butcher-made and factory-made sausages. All factory-made sausages (with the exception of no. 14) lie to the right of the line l, all butcher-made sausages lie to the left of this line. This result was found earlier by van Buuren (1987) who applied MCA to these data for each judge separately, and used Procrustes analysis to match the object scores for each judge.

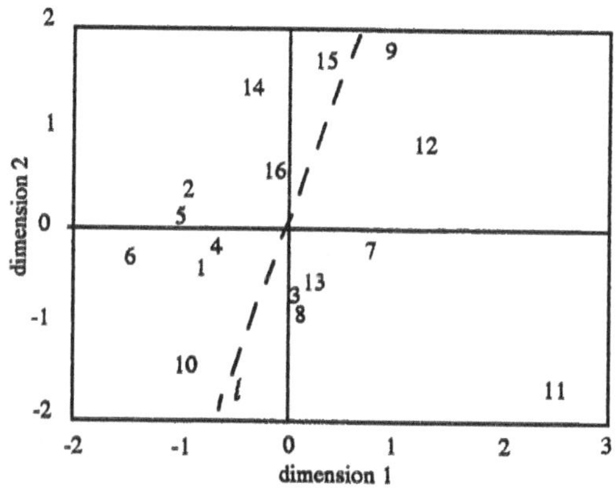

Figure 4 Object scores.

Sausage no. 11 takes a special position. This sausage (factory-made) has been assessed only by judge 3 as having a nice appearance, while the other judges assess this sausage as bad or neutral. Consequently this sausage lies 'far away' in the second quadrant, opposite to the forth quadrant, which corresponds to a nice appearance for most judges (compare Figure 2).

When we compare Figures 2 and 3 with Figure 4 we can see the variables in relation to the objects. From this comparison we can draw two conclusions:

1. Most judges judge butcher-made sausages as more expensive than factory-made sausages.
2. The longest 'nasty taste' vectors point to the direction of the sausages 9, 12 and 15.

When we define a kind of 'overall quality' by high scores on the variables 'appearance', 'taste' and 'odour' and low scores on 'nasty taste', we can conclude that for most judges the 'high quality' sausages lie in the butcher-half of the canonical space, in the third quadrant.

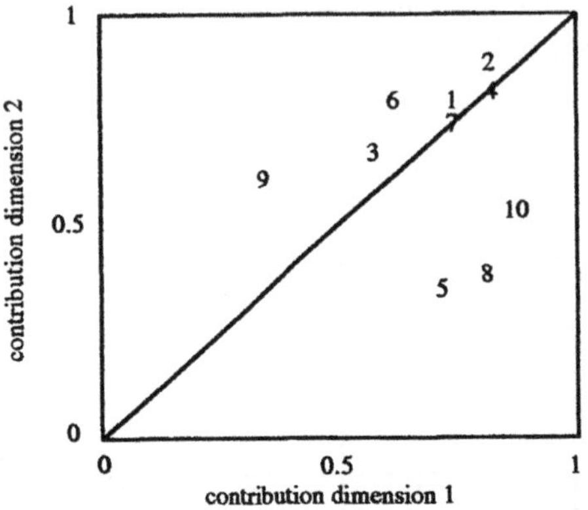

Figure 5 Contributions of the canonical variates for each set.

We are interested in the role of the variables, but also in the role of the judges. Therefore we made a plot of the contributions of each judge to each dimension (Figure 5). This contribution corresponds with the share of each set to the fit. From this plot we can identify outlying judges and clusters of judges that have similar correlations. In Figure 5 we see three judges (5, 8 and 10) for which the first variate has contributed more than the second. Judges 1, 2, 4 and 7 have an equal contribution of the variates and for judges 3, 6 and 9 the second variate was more important.

5.5 Conclusion

We have introduced a form of *k*-sets nonlinear canonical correlation analysis named OVERALS. This technique has been described first by Gifi (1981). A more detailed description is found in van der Burg *et al.* (1988), where OVERALS is treated as a form of *k*-sets homogeneity analysis. Van der Burg and De Leeuw (1987) introduce OVERALS as *k*-sets canonical correlation analysis. In this article both interpretations are found (compare van der Burg 1988, Chapter 4 and 5).

OVERALS is a technique that analyses sets of variables measured on the same objects. The sets of variables are usually not similar. If they are similar, OVERALS allows for different interpretations of the variables. We have seen that in sensory research, which often yields three mode data, OVERALS can be applied. Each judge is represented by a set of variables in the analysis.

Using the OVERALS program we are able to study all three modes at the same time. The variables can be transformed linearly as well as nonlinearly so we can also handle mixed measurement level data.

In the application, we showed that OVERALS can be a useful analysis technique for multiway data with three levels in general, and for sensory data in particular.

CHAPTER 6

Nonlinear Generalised Canonical Analysis: Introduction and Application from Sensory Research

6.1 Introduction

Generalised canonical analysis (abbreviated as GCA) is a technique proposed by Carroll (1968). It is a method for detecting the common features between sets of variables. GCA is a generalisation of canonical correlation analysis (abbreviated as CCA) to more than two sets of variables. CCA, originally proposed by Hotelling (1936), is found in many textbooks on multivariate analysis (e.g. Tatsuoka 1988). Van der Burg, De Leeuw and Verdegaal (1988) suggested a generalisation of Carroll's technique by means of optimal scaling so that data of different measurement levels can be analysed. This method, or rather the computer program that performs this method, is called OVERALS.

Originally published in 1993 in *Advances in longitudinal and multivariate analysis in the behavioural sciences*. Oud, H., van Blokland-Vogelesang, R. (Eds.). (Reprinted with kind permission from ITS)

The generalisation of CCA, which is used by Carroll (1968), is not the only one. In fact there exists a whole variety of generalisations of CCA to a *k-sets* technique. Kettenring (1971), Van de Geer (1984) and Gifi (1990) discuss many possibilities. In this paper we restrict ourselves to the version of Carroll. In addition, we show how the technique of Carroll (1968) and of Van der Burg, De Leeuw and Verdegaal (1988) are related mathematically. Although several articles refer to this relation (Van der Burg 1988, p.99; Van der Burg, De Leeuw and Verdegaal 1988; Gifi 1990, p. 198) it is not shown in detail earlier. Since OVERALS is easily available (SPSS 1990) and more people will use it, it is worth showing this relationship.

In sensory research very often data consist of products, measured on several characteristics or attributes for different assessors. Therefore a type of analysis that can handle sets of variables is relevant in this field of research. As nonlinear GCA has hardly been used in sensory research, we want to show a sensory research application of nonlinear GCA in this article. To this end we perform a secondary analysis on data collected on vegetable soups. The data form a three way table of assessors × products × attributes (c.f. section 6.4). We are interested in answering questions about the (dis)similarity in use of the attributes by the assessors. Because of the common practice to treat ordinal data as numerical data, we are, in addition, interested in the difference between a linear and a nonlinear solution. Although linear GCA is shown to be exactly similar to Carroll's technique, we do not know of a computer program that performs Carroll's technique, so we have to make do with what we have. Using OVERALS we are restricted by the limitations of this computer program, that is, only data without many different scores per variable are accepted. Therefore, we recoded the data from scales with 99 scores into scales with 10 scores, which is a generally accepted practice in exploratory analysis.

Another technique that can handle sets of variables and which is familiar in sensory research is Generalised Procrustes Analysis (GPA, Gower 1975; Arnold and Williams 1985; Dijksterhuis and Gower 1991/2). A nonlinear Procrustes Analysis is proposed by Van Buuren and Dijksterhuis (1988). This nonlinear GPA however, does not estimate so-called isotropic scaling factors. Since the majority of GPA applications in sensory research use these factors, it is hard to compare this nonlinear GPA with linear GPA. Nevertheless, a comparison between linear GPA and linear GCA would be interesting, but is beyond the scope of this paper.

In treating the application, Procrustes Analysis is used for the comparison of linear and nonlinear results from GCA. We provide a reference for Procrustes Analysis without further explanation.

In the following sections we discuss Carroll's technique and show that linear GCA is mathematically related to this method. Next a short introduction to nonlinear GCA is given. The subsequent sections, which form the larger part of the paper, contain a detailed description of the analysis of the vegetable soup data by means of nonlinear GCA.

6.2 Generalised Canonical Analysis

As was mentioned in the introduction, researchers interested in GCA try to answer a question concerning the relationship between sets of variables. Many data sets can be interpreted as consisting of a number of smaller sets. For instance, three-way tables of objects × variables × time-points can be considered as data with a set structure. The data for every time-point form a set of objects × variables. In this case every set has the same number of variables and the variables of each set are similar (have the same interpretations). If different variables are measured on similar objects and the variables are naturally grouped, we also have a set structure. The latter situation is more general than the first one. Let us presume the second situation. Denote a set of variables by $\mathbf{H}_j$ ($n \times m_j$), thus n objects are measured on m_j variables for k sets ($j=1,..,k$). GCA according to Carroll (1968) is defined as the problem of finding linear combinations of the variables in each set so that the squared correlations between the weighted sums and an unmeasured variable $\mathbf{z}$ ($n \times 1$) is maximal. Thus maximise

$$\sigma_1(\mathbf{z},\mathbf{a}) = \sum_{j=1}^{k} \left[\rho(\mathbf{z},\mathbf{H}_j\mathbf{a}_j)\right]^2 \tag{1}$$

where ρ denotes the correlation coefficient, $\mathbf{a}_j$ ($m_j \times 1$) represents the vector of weights for each set and $\mathbf{a}$ refers to the vector containing all $\mathbf{a}_j$. In fact, Carroll (1968) uses an even more general function, as he gives each correlation a positive weight. However, we are not interested in these weights so they are left out. The fit measure σ_1 (denoted as R^2 by Carroll) can vary from zero to k. The solution for the $\mathbf{a}_j$ is the well known regression solution. The weighted linear combination $\mathbf{H}_j\mathbf{a}_j$ can be expressed in terms of the projector of $\mathbf{H}_j$, that is

$$\mathbf{H}_j\mathbf{a}_j = \mathbf{H}_j(\mathbf{H}'_j\mathbf{H}_j)^{-1}\mathbf{H}'_j\mathbf{z} = \mathbf{P}'_j\mathbf{z} \text{ with } \mathbf{P}_j = \mathbf{H}_j(\mathbf{H}'_j\mathbf{H}_j)^{-1}\mathbf{H}'_j \tag{2}$$

If $\mathbf{H}_j$ is not of full rank the generalised inverse can be used. Carroll's method can now be rewritten in terms of $\mathbf{P}_j$, which gives for the fit

$$\sigma_1(\mathbf{z}, *) = \sum_{j=1}^{k} (\rho(\mathbf{z},\mathbf{P}_j\mathbf{z}))^2 = \sum_{j=1}^{k} \frac{(\mathbf{z}'\mathbf{P}_j\mathbf{z})^2}{(\mathbf{z}'\mathbf{z})(\mathbf{z}'\mathbf{P}_j\mathbf{z})} = \sum_{j=1}^{k} \frac{\mathbf{z}'\mathbf{P}_j\mathbf{z}}{\mathbf{z}'\mathbf{z}} \tag{3}$$

The * in the fit shows that the maximum is taken over the corresponding unknowns. Using $\mathbf{P}$ for $k^{-1}\sum_{j=1}^{k}\mathbf{P}_j$ the average projector, expression 3 changes into

$$\sigma_1(\mathbf{Z}, *) = k\frac{\mathbf{z}'\mathbf{P}\mathbf{z}}{\mathbf{z}'\mathbf{z}} \tag{4}$$

Maximising expression 4 comes down to the eigenvalue decomposition of $\mathbf{P}$, where $\mathbf{z}$ is proportional to the first eigenvector and the maximum corresponds with k times the first eigenvalue. A second solution is found by taking the second eigenvector, and so on. Another way of describing the GCA problem for p solutions is to maximise

$$\sigma_1(\mathbf{Z}, *) = \mathrm{tr}(k\mathbf{Z}'\mathbf{P}\mathbf{Z}) \text{ subject to the condition that } \mathbf{Z}'\mathbf{Z} = \mathbf{I} \tag{5}$$

where matrix $\mathbf{I}$ is the $(n \times n)$ identity matrix and $\mathbf{Z}$ is an $(n \times p)$ matrix of unmeasured variables. The maximum is proportional to the sum of the eigenvalues of $\mathbf{P}$, and $\mathbf{Z}$ corresponds to the first p eigenvectors of $\mathbf{P}$. Expression 5 is closely related to the generalisation of GCA which is proposed by Van der Burg (1988, chapter 4 and 5); Van der Burg, De Leeuw and Verdegaal (1988), and Gifi (1990, chapter 5). These authors minimise

$$\sigma_2(\mathbf{X},A) = (nk)^{-1}\sum_{j=1}^{k} \mathrm{tr}(\mathbf{X}-\mathbf{H}_j\mathbf{A}_j)'(\mathbf{X}-\mathbf{H}_j\mathbf{A}_j) \tag{6}$$

subject to the conditions that $\mathbf{X}'\mathbf{X} = n\mathbf{I}$ and $\mathbf{u}'\mathbf{X} = \mathbf{0}$

where $\mathbf{u}$ is an n-vector of ones and $\mathbf{0}$ an n-vector of zeros. The matrix $\mathbf{A}$ consists of all $\mathbf{A}_j$ written under each other. The solution for $\mathbf{A}_j$ $(m_j \times p)$ is the p-dimensional regression solution. Thus we get

$$\mathbf{H}_j\mathbf{A}_j=\mathbf{H}_j(\mathbf{H}'_j\mathbf{H}_j)^{-1}\mathbf{H}'_j\mathbf{X}=\mathbf{P}_j\mathbf{X}$$

Substitution of this in expression 6 gives:

$$\begin{aligned}\sigma_2(\mathbf{X}, *) &= (nk)^{-1}\sum_{j=1}^{k}\mathrm{tr}(\mathbf{X}-\mathbf{P}_j\mathbf{X}_j)'(\mathbf{X}-\mathbf{P}_j\mathbf{X}_j)\\ &= (nk)^{-1}\sum_{j=1}^{k}\mathrm{tr}(\mathbf{X}'\mathbf{X}-\mathbf{X}'\mathbf{P}_j\mathbf{X}_j)'(\mathbf{X}-\mathbf{P}_j\mathbf{X}_j)\\ &= p-n^{-1}\mathrm{tr}X'\mathbf{P}\mathbf{X}\end{aligned} \tag{7}$$

subject to the conditions that $\mathbf{X}'\mathbf{X} = n\mathbf{I}$ and $\mathbf{u}'\mathbf{X} = \mathbf{0}$

Using the method of Lagrange multipliers for minimisation of expression 8 shows that **X** is an orthogonal rotation of the eigenvectors of **P** (except the first eigenvector) and that the minimum is p minus the sum of the corresponding eigenvalues of **P** (Van der Burg, De Leeuw and Verdegaal 1988). If the matrices $\mathbf{H}_j$ consist of standardised scores, which is the case in Van der Burg, De Leeuw and Verdegaal (1988), the first eigenvector is automatically removed from **P**. Thus we see that Carroll's method and expression 8 are essentially equivalent.

6.3 Nonlinear Generalised Canonical Analysis

A nonlinear version of generalised canonical analysis can be obtained by introducing optimal scaling (Young 1981). The scores for the variables (columns of $\mathbf{H}_j$) are replaced by transformed variables (columns of $\mathbf{Q}_j$, where $\mathbf{H}_j$ and $\mathbf{Q}_j$ are of the same order) which satisfy the measurement restrictions. The method of optimal scaling solves for the various possible transformations by minimising the criterion function, which, in our case, is defined in equation 8. There are three types of measurement restrictions: nominal, ordinal and numerical measurement restrictions (interval level). In the nominal case the ties (similar scores) per variable remain tied in the transformation. In the ordinal case the ties are kept and the order of the scores is maintained too. In the numerical case only linear transformations of the original scores are allowed, which implies that $\mathbf{Q}_j$ equals $\mathbf{H}_j$ as both matrices consist of standardised scores. However, for the sake of notation it is easier to see $\mathbf{Q}_j$ in the numerical case also as a transformation matrix. Let us denote a column of $\mathbf{H}_j$ by $\mathbf{h}_{js}$ and a column of $\mathbf{Q}_j$ by $\mathbf{q}_{js}$ ($s=1,\ldots,m_j$) and symbolise the transformations that satisfy the measurement restrictions by $C(\mathbf{h}_{js})$. Then GCA with optimal scaling corresponds with minimising

$$\sigma_3(\mathbf{X},\mathbf{Q},\mathbf{A})=(nk)^{-1}\sum_{j=1}^{k}\operatorname{tr}(\mathbf{X}-\mathbf{Q}_j\mathbf{A}_j)'(\mathbf{X}-\mathbf{Q}_j\mathbf{A}_j) \tag{9}$$

subject to the conditions that $\mathbf{X}'X=n\mathbf{I},\mathbf{u}'\mathbf{X}=\mathbf{0}$ and $\mathbf{q}_{js}\epsilon C(\mathbf{h}_{js}),s=1,\ldots m_j,j=1,\ldots,k$

where $\mathbf{Q}$ denotes the matrix of all $\mathbf{Q}_j$ written next to each other. Expression 9 can also be provided with the option of so-called multiple nominal quantifications. This relates nonlinear GCA to multiple correspondence analysis (Nishisato 1980; Greenacre 1984). However, we do not need this for the application we will discuss in the next section, therefore we do not add this type of transformation. Nonlinear GCA as defined by expression 9 and extended with multiple nominal transformations is called OVERALS by Van der Burg, De Leeuw and Verdegaal (1988). In fact the computer program that realises the method is called OVERALS. This program is available as an SPSS module (SPSS 1990).

Note that the solution for the $\mathbf{A}_j$ is again the regression solution. However, this time the solution depends on the $\mathbf{Q}_j$, which are unknowns. Therefore we write $\mathbf{P}(\mathbf{Q})$ to show that $\mathbf{P}$ is no longer fixed. The solution for $\mathbf{X}$ is, in this case, proportional to an orthogonal rotation of the eigenvector matrix of $\mathbf{P}(\mathbf{Q})$ which shows the dependence on the transformations. If all variables are numerical, all $\mathbf{Q}_j$ are equal to $\mathbf{H}_j$ and we are back in formulation 6.

The eigenvalues of $\mathbf{P}(\mathbf{Q})$ are a fit measure for every solution or dimension. They vary between zero and one and their sum is p minus the loss (see expression 9). The sum of eigenvalues varies between zero and p.

The matrix $\mathbf{X}$ is called the matrix of object scores. The correlations between the transformed variables ($\mathbf{q}_{js}$) and the object scores are called the *component loadings* in analogy with principal component analysis (PCA). The component loadings represent a point in the $\mathbf{X}$-space for each variable. If every point is connected with the origin we get a plot of vectors representing the variables, which can be interpreted in a similar way as in PCA (the difference with PCA is the way the $\mathbf{X}$-space is obtained). In addition, there are scores for each object (the rows of $\mathbf{X}$), so that both objects and variables can be plotted. We will use these plots to interpret the solution of the application discussed in the next section.

6.4 Application from Sensory Research

The data to analyse are from sensory research. Forty vegetable soups were rated by nineteen trained assessors on five variables concerning odour, taste and mouthfeel attributes (see Table 1). The scores vary from 0 to 98, the assessors used a line scale to indicate their ratings. A score of 0 indicates that an attribute was not present according to the assessor, a score of 98 means that the attribute is very strongly present in a particular soup. Apart from the attributes, the brand of the soup, the package and the type are known, (see Table 1). In this application the odour, taste and mouthfeel attributes are analysed. They form a $40 \times 19 \times 5$ multiway table. The other characteristics are used to interpret the solutions and to identify groups.

Originally the vegetable soup data were gathered as part of a larger experiment in which chemical and microscopical data also were collected (Cramwinckel and Van Mazijk-Bokslag 1990). The object of that experiment was to get information about the contents, the sensory quality and the taste of forty different vegetable soups. The main objective of the secondary analysis reported in this paper is not the quality or taste of the soups but the use of the five attributes by the assessors. It is interesting to know on which attribute(s) the assessors mostly agree and which attributes give rise to confusion. In addition, it is interesting to find out whether the partition of the soups, according to the ratings, makes sense in the light of the type or package of the soup. In order to try to answer these questions, the ratings are analysed with OVERALS, using the assessors as sets, the soups as objects and the attributes as variables. Comparison of a solution with ordinal measurement levels and one with numerical measurement levels also enables us to investigate whether this kind of data should be analysed under metric or non-metric assumptions. In the following paragraphs, we show that we use three categories for the ordinal transformations and ten for the numerical transformations, and we also explain why this is done.

Two assessors (sets) were removed from the data because they had too many missing values, leaving 17 sets. The computer program OVERALS can only take a relatively small number of different scores per variable, therefore the data were categorised into a small number of categories. Some preliminary ordinal analyses with different numbers of categories (seven, six and five were tried) showed an extreme outlier in a three-dimensional solution. Removal of the outlier resulted in a new outlier, removal of this outlier again showed a new outlier. The same was encountered with a two-dimensional solution. This phenomenon seemed due to the fact that many of the higher scores and some of the lower ones occurred only once. To cope with the outliers the number of categories was reduced to just three, resulting in less extreme outliers (they did not vanish). This data reduction appears not too severe, as the numerical ten-category and the ordinal three-category analyses show comparable and interpretable results (c.f. section 5). In a previous OVERALS-analysis, a partition of the variables into three categories also resulted in an interpretable solution (Van der Burg and Dijksterhuis 1989). The three categories used for all attributes, are 1 (scores 0 to 25), 2 (scores 26 to 50) and 3 (scores 51 to 98). The third category was taken larger than the first and second category to take account of the skewness. For four variables, one category was rescaled to a lower or higher value, as these four categories still contained only one observation.

Table 1 Variables in the Vegetable Soup Research

Odour and taste attributes	
Spiciness	(0 = not, ..., 98 = much)
Vegetables	(0 = not, ..., 98 = much)
Saltiness	(0 = not, ..., 98 = much)
Mouthfeel attributes	
Thickness, jelly-like	(0 = not, ..., 98 = much)
Firmness of vegetables	(0 = not ,...., 98 = much)
Package	concentrated in tin
	instant in packet
	dried in packet
	deep-frozen
	ready-to-eat in tin
Type	ordinary
	cream
	chinese
Brand	brand-names

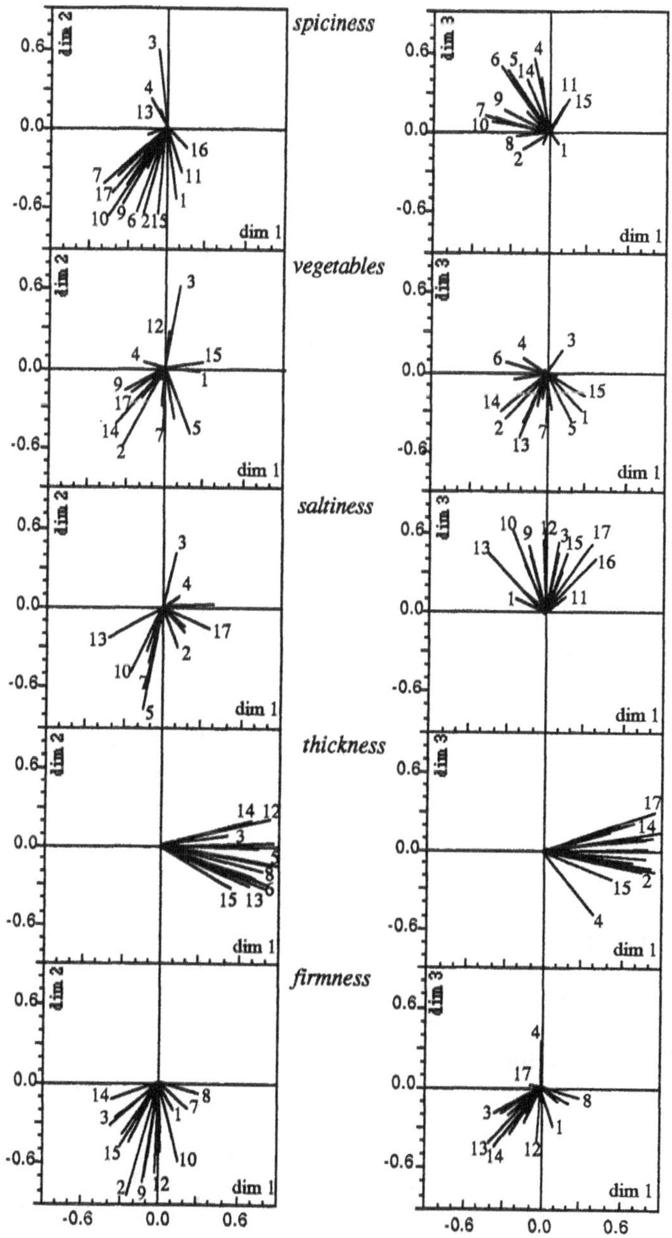

Figure 1 Component Loadings of the Ordinal Solution for Each Attribute Separately

Table 2 Eigenvalues of a Three-dimensional Solution with Ordinal and Numerical Measurement Levels

Analysis	No of categories	EIGENVALUES			SUM
		dim.1	dim.2	dim.3	
Ordinal	3	.684	.533	.433	1.650
Numerical	10	.715	.510	.429	1.654

To perform a numerical analysis (i.e. with only numerical measurement levels), we were restricted by the possibilities of the OVERALS computer program, as 99 scores for 17 × 5 variables are too many for the computer program. However, a larger number of categories was used than for the ordinal analysis. The scores were recoded into ten categories, corresponding to ten equal sized intervals (category 1 consists of scores 0 to 9, 2 of scores 10 to 19, ..., 10 of scores 90 to 99). The last category was empty for every variable. Of course, recoded data are never similar to the original data, implying that $\mathbf{Q}_j$ is not equal to $\mathbf{H}_j$. However, it is our experience that usually hardly any information is lost. The difference between a solution with many categories (about 10) and very many categories (more than 20) is usually nil for multivariate techniques. Thus the numerical ten-category analysis is not a real numerical analysis but a quasi-numerical analysis that is likely to give very similar results.

6.5 Results

A three-dimensional solution is obtained in both the ordinal three-category and the numerical ten-category analysis. The eigenvalues of the three-dimensional solutions were not really high (see Table 2). The maximum of an eigenvalue is one and the minimum zero. A sum of approximately 1.7 indicates that the solutions found are not very strong. In general, high eigenvalues are needed to correspond with much variance in each set. This gives an indication that the differences between the sets are probably due to individual differences in the use of the five attributes. As can be seen from Table 2, the two solutions have rather similar eigenvalues. It appears that the ordinal GCA solution based on just three categories is about as good as a numerical GCA solution based on ten categories. A high number of categories does perhaps provide only superfluous information. To see if this conclusion holds, not only are the eigenvalues compared but also both the component loadings and the object scores are matched.

In order to see how the attributes were used by the assessors, plots of the component loadings are given for each attribute separately. Thus, five plots are made instead of one. Dimension 1 and 2, and dimension 1 and 3 are plotted against each other, resulting in 10 separate plots. In Figure 1 these plots are shown for the ordinal solution. Not all vectors are identified by their set number. However, for interpretation of the figure this will do.

A Procrustes rotation (see e.g. Gower 1975) to match the three-dimensional configurations of the component loadings for the ordinal three-category and the numerical ten-category solution, revealed a virtual identity of solutions: 97% of the variance of the two configurations was matched. For this reason, the results of the numerical ten-category solution are not shown. The high match confirmed the idea that the ordinal three-category and numerical ten-category solution provide the same information.

Inspection of Figure 1 shows that the assessors agree very much on the attribute 'thickness'. The first dimension depends mostly on this attribute. The second dimension is dominated by 'spiciness' and 'firmness'. Assessor 3 and, to a lesser extent, assessor 4 are noted as persons who do not agree with the other assessors about 'spiciness'. The variables of 'vegetables' correlate mostly with the second dimension, although the agreement between the assessors is much less on this attribute than on 'firmness' or 'spiciness'. With respect to 'vegetables' assessors 3 and 12 behave rather exceptionally. The third dimension is, for the larger part, determined by 'saltiness'.

Apparently 'thickness' is the attribute on which most assessors agree, followed by 'spiciness' and 'firmness', and finally by 'saltiness'. It seems that 'vegetables' is one of the most difficult attributes to agree on. Especially assessor 3 must be mentioned as a person who interprets many attributes differently from the other assessors.

In Figure 2 the object scores of the ordinal solution are shown. The soups are labelled according to package and type. To check if the ordinal three-category and the numerical ten-category solutions are alike, again a matching is performed with the help of a Procrustes rotation. This rotation accounts for 98% of the variance in the solutions, so only 2% is lost in the matching process. Therefore the solutions are almost identical. Again only the ordinal configuration is shown.

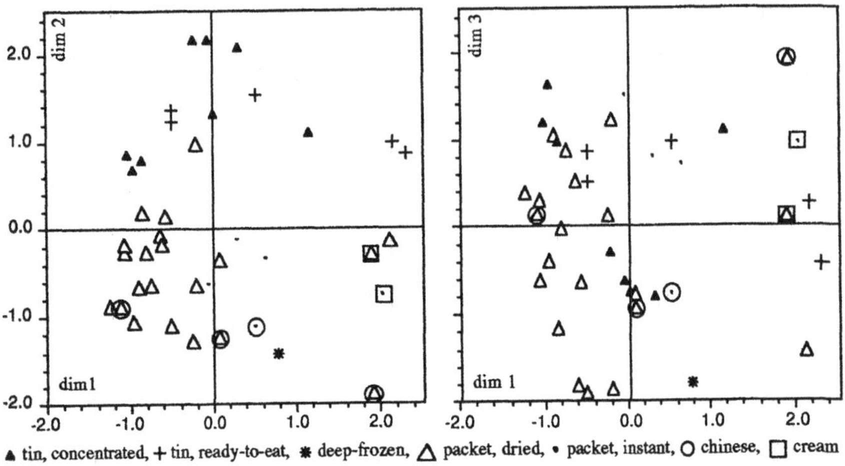

Figure 2 Object Scores of the Ordinal Solution, Labelled by Type and Package.

From Figure 2 it can be inferred that there does exist an overall difference in taste between soup from tins and from packets (and deep-frozen). This distinction is found along the second dimension, with the tins having higher scores than the packets. Since 'spiciness' and 'firmness' were the attributes associated mostly with the second dimension (in negative direction), it seems that soups in tins are less spicy and firm than in packets.

Most of the soups have low scores on the first dimension. The six soups that have higher scores, seem thicker than other soups. As may be expected, the two cream soups are found among the thicker types. The third dimension is not directly interpretable in terms of package or type of soup. Figure 1 showed that 'saltiness' is the main attribute for the third dimension, so it can be concluded that the judged 'saltiness' of the soups has no clear relation with the package of the soup. The four 'Chinese' vegetable soups all have low scores on the second dimension, so they do possess firm vegetables and are rather spicy. In other respects the Chinese soups do not differ much from the other soups. The one deep-frozen soup also seems to have a non-salty taste and firm and spicy vegetables.

Labelling the plot of object scores by brand did not show any regularity. Therefore, the names of the soups are not shown in Figure 2.

6.6 Conclusion

By analysing *k*-sets data from the field of sensory research with OVERALS, a lot of aspects of the data can be studied. Plotting the component loadings of a variable for all sets, as in Figure 1, provides a useful way to identify the consistency of the use of the variables (in this case attributes) by the sets (assessors in the panel). The relative position of the objects (vegetable soups) shows the more important (dis)similarities between the objects. The object scores can be compared with the component loadings to see which variables are responsible for congruencies or differences between them. In addition, labelling the object scores by external variables (package, type), variables not used in the analysis, also helps to interpret the solution.

It can be concluded from the analysis that it is not useful to analyse a large number of categories. The results from an ordinal analysis with only three categories and the results from a numerical analysis with ten categories are almost identical. Apparently the higher number of categories does not provide much extra information. An ordinal ten-category solution has not been compared with the numerical ten-category solution, as this solution will capitalise on unique patterns in the data.

Van der Burg and De Leeuw (1988) discuss the stability of OVERALS solutions. They perform several jackknife and bootstrap studies. These studies are not replicated in this application although it would be worthwhile. We give it as an option for other users of the OVERALS program.

As mentioned in the introduction, GPA is another technique for *k*-sets analysis. This technique is familiar in sensory research. Therefore a comparison between results from GCA and GPA applied to the same data will be interesting.

Concluding Remarks Part II: Measurement Levels

Summary

The findings in the three chapters in this part are summarised and commented upon. In addition some suggestions for further research are given.

Chapter 4
Multivariate Analysis of Coffee Images

This chapter presented a method that enables the analysis of both categorical and quantitative variables to be carried out using a GPA. The categorical variables are first analysed by means of Multiple Correspondence Analysis and the quantitative variables by a PCO. They are subsequently combined into individual matrices which can be matched by means of GPA. In fact they can be matched in a number of ways, stressing the objects, the categories, the quantitative variables, or combinations of these. Strictly speaking, one is free to use a method other than GPA, e.g. GCA, though standard GPA, matching the objects, is the obvious choice. Putting the stress on the fit of categories could perhaps be useful in special applications. Matching the quantitative variables introduces a number of possibilities, briefly touched upon in the chapter, but not employed further.

The choice for MCA, PCO and GPA is not a necessary restriction. A lot of different methods can be used within the framework sketched. The method is very flexible and allows for the combination of different distance-models and multivariate analyses of the individual assessor's variables.

Though seven assessors is a rather low number, the data serve as an illustration of the possibilities of the method. Inspecting the individual biplots of the nine coffee-brands, the categories of the qualitative variables and the positions of the quantitative variables enables a detailed study of the structure in the data. Matching the individual configurations, using e.g. standard GPA, retains the most salient features in the data.

Chapter 5
Nonlinear Canonical Correlation Analysis of Multiway Data

Ordinal GCA can be used on data with only three categories and can produce an interpretable result. Of course information is lost in the recoding of the numerical data into only three categories. A comparison of other categorisations would be useful, but is wanting. Another interesting analysis is the ordinal, or even nominal analysis of the ranked data. The scores of the variables can be replaced by rankings. These rankings can be analysed by OVERALS. Such an analysis of the rankings would bypass problems in the recoding of the original 50 categories into the rather low number of three categories.

A meaningful configuration of sausages, showing a clear division between butcher made and factory made sausages emerged. The object scores (Chapter 5, Figure 4) as well as the component loading plots (Chapter 5, Figure 1 and Figure 2) can be inspected. The modelling of nonlinearities can be studied in the plots of the category quantifications (Chapter 5, Figure 3). The distinction between butcher-made and factory-made sausages was obtained by van Buuren (1987) too. Van Buuren applied Multiple Correspondence Analysis to each individual data set and matched the results using GPA. Note that this is an application of the general scheme presented in Chapter 4 (see formula (1) in the Introduction to Part II and in Chapter 13, §13.3.1. The Group Average configuration van Buuren presents (1987, Fig. 2, p.449) shows a clearer grouping of the sausages. There are a butcher-made and a factory-made cluster of sausages and some sausages in between. The less clear grouping of the GCA result may be a result of a difference in data or in method (or both). Van Buuren used all 27 attributes of the original data while in Chapter 5 only 5 of these variables were used. Matching MCA results, compared with matching the original individual data sets, may result in loss of fit because each set's MCA solution is constructed irrespective of the other sets. It could happen that there is potential fit excluded in this way, especially since GCA is sensitive to common directions in the sets. As a result the configuration of sausages obtained by GCA can be more homogeneous than that obtained by the MCA-GPA method of van Buuren because there were more "fitting" directions in the individual spaces.

The plot with the contributions of assessors to the dimensions (Figure 5) is a plot not found in other literature on GCA. This plot was taken from Dijksterhuis (1987)[1] and is based on the losses per dimensions for the individual assessors. This loss per dimension is made into a kind of fit by subtracting it from one. It results in an Indscal-like individual weights plot. The plot illustrates the importance of a particular dimension for an assessor.

Around the time the research for this chapter took place, no other application of GCA to a 3-way data set, assigning each individual set an assessor, was encountered in the literature (Dijksterhuis 1987). The analysis of sensory data, as illustrated in this chapter, turned out to be an interesting new application area of GCA.

Chapter 6
Nonlinear Generalised Canonical Analysis: Introduction and Application from Sensory Research

Comparing an ordinal analysis of a very low number of categories with a numerical analysis of a larger number of categories, recoded from the same data set, is a useful enterprise. The reason to recode the original data into a rather low number of categories was to try to remove unique patterns giving rise to clear outliers in the result. The larger the number of categories, the more unique categories will arise. GCA is sensitive to such unique categories, so the data were recoded into the minimal number of categories that allowed an ordinal analysis: three. The 10-category numerical analysis was carried out to check whether the recoding into three categories had not resulted in a great loss of information.

Table 1 summarises the very small differences found after Procrustes matching the two resulting configurations. Both the configurations with the objects scores and with the component loadings were separately matched.

[1]It was suggested by Stef van Buuren.

Table 1 Agreement of the object scores and the component loadings of a 3-category ordinal with a 10-category numerical GCA solution.

Configuration	Figure	Procrustes match (variance lost)
Component loadings	Chapter 6, Figure 1	3%
Object scores	Chapter 6, Figure 2	2%

That the resulting two configurations show such high agreement could lead to two conclusions:

1. there is no need for line-scales, just give the assessors a low number of categories to choose from

2. there is no need for nonlinear multivariate methods, just perform linear analyses of line-scale data.

Both conclusions however are probably premature. Additional research is needed as will be suggested in the concluding chapter (Chapter 13).

PART III

SENSORY-INSTRUMENTAL RELATIONS

INTRODUCTION TO PART III

Summary

In this part three different multivariate statistical methods to study Sensory-Instrumental relations are introduced:

1. (nonlinear) Redundancy Analysis (Chapter 7 and 8)
2. (nonlinear) Canonical Correlation Analysis (Chapter 8)
3. Procrustes Analysis (Chapter 9)

Nonlinear and linear analyses are compared for Redundancy Analysis and Canonical Correlation Analysis. Parts of the same data set are used in all three chapters.

Introduction

Figure 1 illustrates the structure of the data used in the three chapters in this part. For two varieties of apples, Cox and Elstar, 72 apples are judged by three trained assessors on the attributes mealyness and firmness. The apples were also subjected to a number of instrumental measurements, which are briefly explained in the Chapters 7, 8 and 9.

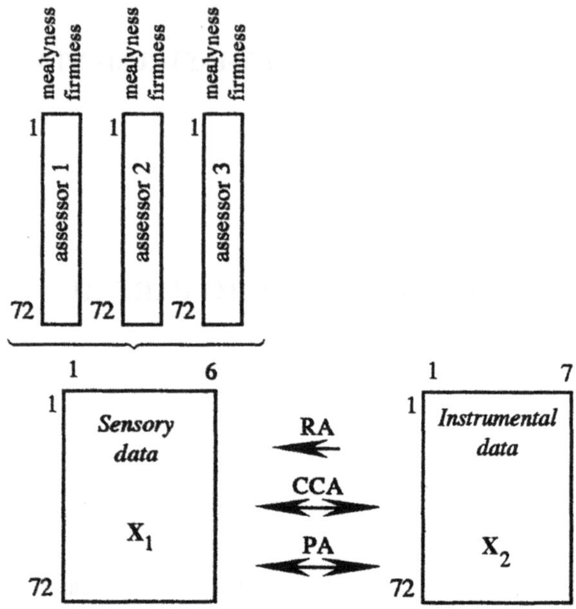

Figure 1 Structure of the apple data in the analysis of the Sensory-Instrumental data sets. The arrow pointing to the left represents the asymmetric analysis: Redundancy Analysis (RA), the double arrows represent the symmetric analyses: Canonical Correlation Analysis (CCA) and Procrustes Analysis (PA).

The differences between the three methods will be summarised in the concluding remarks at the end of Part III.

It was argued in Chapter 1 (Introduction) that non-linear analyses might be useful in the analysis of Sensory-Instrumental relations. Nonlinearity may be encountered especially when appreciative attributes are involved (§1.7.3), but textural attributes in the apple data set may also give rise to nonlinearities.

Though the three chapters analyse -parts of- the same data set, they were originally not intended for the comparison of the three MVA methods. As a consequence it may be hard to compare the methods and to keep track of which part of the apple data is analysed by which methods. The Concluding Remarks of this part give an overview of the analyses and the results.

Chapter 7
An Application of Nonlinear Redundancy Analysis[1]

In the first part of this chapter, Redundancy Analysis is briefly introduced. The extension of linear Redundancy Analysis to include non-linear transformations of the variables is introduced using optimal scaling. Nonlinear Redundancy Analysis, also known by the name of the corresponding algorithm, "Redundals", was developed by van der Burg and de Leeuw (1990). Because it is a two-data-set method where one of the sets is predicted from the other set, and because nominal and ordinal variables can be analysed, the method was thought apt for the problem of the apple-data sets analysed in this chapter. The apple data contain some background variables (nominal and ordinal), instrumental variables (numerical) as well as sensory judgements (ordinal, maybe numerical) on two varieties of apples. Both a linear and a nonlinear Redundancy analysis are performed and their results compared.

In the chapter there is mention of a Procrustes rotation to match the solutions from the ordinal and the numerical Redundals analysis. The percentage of 97.5% variance accounted for (for the Cox-data), means that after rotating the two configurations towards each other, only 100%-97.5% = 2.5% variance would be lost in subsequent averaging the two configurations. This means that the two configurations are very similar. Of course there is no point in averaging the numerical and the ordinal solution here, it is a way to illustrate the concept of Variance accounted for -and lost- in the context of Procrustes matching two configurations from different analyses.

[1] The paper was presented by Eeke van der Burg to illustrate the Redundals technique in a presentation at the *International Workshop Multidimensional Data Analysis. Meeting of Dutch & Italian Schools*, held in Anacapri, Italy, 2 to 5 October 1991. It was printed in a special issue of Statistica Applicata, the Italian Journal of Applied Statistics (1992, vol. 4, no. 4, p. 565–575). There is no computer program for Redundals available, the program used for the analyses in the chapter was an experimental FORTRAN program written by Eeke van der Burg.

Chapter 8
An Application of Nonlinear Redundancy Analysis and Canonical Correlation Analysis[2]

This chapter contains a short paper in which both Nonlinear Redundancy Analysis and Nonlinear Canonical Correlation Analysis (Canals) are applied. The methods are introduced in a very condensed way. The previous chapter contains a somewhat more elaborate introduction of Nonlinear Redundancy Analysis. For Nonlinear Canonical Correlation Analysis the reader is referred to van der Burg and de Leeuw (1983).

There is some overlap between this and the previous chapter. The Redundals results in Table 2 of Chapter 7 are summarised in Table 2 of the current Chapter. Figure 1 of this chapter is identical to Figure 1a of Chapter 7, it shows the correlations of the variables with the axes of the two-dimensional ordinal Redundals solution. In addition to the Redundals solution an ordinal Canals solution is computed. The difference is that with the Canals analyses the background variables were related to the sensory variables and to the instrumental variables separately. The background variables are design variables coding origin, picking season, size-class and storage temperature of the apples. These variables were not used in the Redundancy Analyses in Chapter 7.

[2] The paper was presented by the first author at the 7th European Meeting of the Psychometric Society in Trier, Germany, July 29-31, 1991. It is printed in *Psychometric Methodology, Proceedings of the 7th European Meeting of the Psychometric Society in Trier* (Steyer, Wender, Widaman (Eds.) 1993, p. 74–79).

Chapter 9
Procrustes Analysis in Studying Sensory-Instrumental Relations[3]

In this paper the relation between a sensory and an instrumental data set is studied by means of Procrustes Analysis. The method is introduced with emphasis on its application to match two data sets. First the structure of each data set is studied separately by means of Principal Component Analyses. After standardising the two data sets Procrustes Analysis is used to match the two sets. It is concluded that, though not often used to this end, Procrustes Analysis is a suitable method to study the relations between sensory and instrumental data.

The possible advantage that Procrustes Analysis has over Redundancy Analysis and Canonical Correlation Analysis is the rigidity of its transformations. Especially with CCA it sometimes happens that almost perfect fit is obtained, which can be the artificial result of a very high correlation between only two variables, one in each set. When optimal scaling is included in RA and CCA there is even more freedom to obtain almost perfect fit, which may or may not be an artefact. Perfect fit often poses problems of interpretation.

---

[3] This chapter is based on the presentation given at the "Understanding Flavour Quality" Conference held September 20-23, 1992 at the University of Bristol, England. This conference was organised jointly by ELSEVIER science publishers, Sensory Research Laboratories Ltd. (Nailsea, UK) and Oliemans Punter & Partners (Utrecht, the Netherlands), see Piggott (1993) for an impression of the conference. The paper is printed in Food Quality and Preference (1994, vol. 5, numbers 1 & 2, p. 115–126) in a special issue on the "Understanding Flavour Quality: Relating Sensory to Chemical and Physical Data" symposium.
The two apple data sets were made available by ATO-DLO (Institute for Agrotechnology, Wageningen, the Netherlands).

CHAPTER 7

An Application of Nonlinear Redundancy Analysis

7.1 Introduction

The reason for this study is that we are interested in applications of nonlinear redundancy analysis or REDUNDALS as defined by Van der Burg and De Leeuw (1990). Redundancy analysis is a technique that can be used when two sets of variables are present. The aim of the analysis is prediction of one set from the other in an optimal way. Many types of data can be used for redundancy analysis. However, as REDUNDALS has the possibility to analyse data of a nominal, ordinal or numerical measurement level, we preferred data of a nominal or ordinal level. The apple data analysed by us were collected to search for variables which can predict certain sensory qualities of the apples, i.e. mealiness and firmness. The data were analysed with multiple regression techniques by Koppenaal (1991). In the present paper secondary analyses on these data are performed by means of the REDUNDALS technique. The results of this study agree with the earlier results of Koppenaal (1991). We also

By Eeke van der Burg and Garmt Dijksterhuis. Originally published in 1991 in *Statistica Applicata, Italian Journal of Applied Statistics*, 4, 65-575.

compare the results of a linear REDUNDALS solution with a nonlinear solution using a Procrustes rotation (Cliff 1966, Gower 1975).

In the following sections first an explanation of (nonlinear) redundancy analysis is given. Subsequently an extensive description of the REDUNDALS-analyses of the apple data is provided.

7.2 Redundancy Analysis

Redundancy analysis is a technique named by van den Wollenberg (1977). He introduced it as an alternative for Canonical Correlation Analysis (CCA). In CCA two sets of variables optimally predict each other using linear combinations of the variables per set. This formulation is symmetrical in the way the sets are treated. However often researchers do not consider their sets as symmetrical. Many times one is interested in how well one of the sets can be predicted from the other while explaining a maximum of variance of the criterion set. Let us use the symbol **X** for the criterion set. Suppose **X** contains n rows (objects, individuals, or units) and m_1 columns (variables, measurements or qualities). Let **Y** $(n \times m_2)$ be a set of m_2 predictors measured on n objects. If the predictors are weighted such that every criterion variable is maximally predicted we can formulate this in a least squares way as follows:

$$\text{minimise SSQ}(\mathbf{X}\text{-}\mathbf{YB}) \qquad (1)$$

With **B** $(m_2 \times m_1)$ the matrix of weights and SSQ the sum of squares, SSQ(**A**)=trace **A'A**. Formulation (1) corresponds with multivariate multiple regression. When we presume that the space spanned by the predictors contains a subspace of low rank that predicts the criterion variables, we restrict the weight matrix **B** to a matrix of low rank. This is equivalent with saying that matrix **B** can be split into **V** $(m_2 \times p)$ and **W** $(m_1 \times p)$ with p small ($p \leq \min(m_1,m_2)$). Then we get

$$\text{minimise SSQ } (\mathbf{X} - \mathbf{YVW}') \qquad (2)$$

In the above formulation **YV** corresponds with the low dimensional subspace of the predictors. The **W'** are weights for the linear combinations of **Y**. In (2) there is a lot of freedom in choosing **V** and **W** as postmultiplication of **V** and **W** by a rotation matrix **H** $(p \times p)$ with **HH'**=**I** leads to the same solution. This indeterminacy is common in many multivariate techniques, e.g. in canonical correlation analysis and principal component analysis. Due to the freedom the most convenient **V** can be chosen. Let us take the matrix **V** such that **YV** is a p-dimensional orthogonal basis in the **Y**-space. Thus

$$\mathbf{V'Y'YV} = n\mathbf{I} \tag{3}$$

With **I** ($p \times p$) the identity matrix. The technique defined by (2) and (3) is called redundancy analysis (van den Wollenberg 1977, Israëls 1984 and 1986). This technique is also called reduced rank regression (Izenman 1975, Davies and Tso 1982, ten Berge 1985). A similar technique has been discussed earlier by Rao (1964). See de Leeuw (1986) for a brief history of redundancy analysis.

In (2) no restriction is made on the data sets **X** and **Y**. Restriction to standardised variables is of no importance for the predictor variables, it only changes the weights **V** but not the product **YV**. However the **X** set changes with standardisation. Israëls (1984) describes both possibilities. We restrict ourselves to standardised variables so that $\mathbf{X'Y}/n$ and $\mathbf{X'X}/n$ and $\mathbf{Y'Y}/n$ represent correlation matrices. The case with nonstandardised criterion variables is easily generalised from this paper. Standardisation is formulated as

$$\mathbf{u'x} = 0,\ \mathbf{u'y} = 0 \text{ and } \mathbf{x'x} = n,\ \mathbf{y'y} = n \tag{4}$$

With **x** a column of **X** and **y** a column of **Y** and **u** an n-vector of ones.

7.3 Optimal Scaling

Van der Burg and De Leeuw (1990) use optimal scaling for their variables to handle data of which nominal, ordinal or numerical measurement levels are assumed. Optimal scaling (Young, 1981) is defined as nonlinear transformations of the variables such that measurement restrictions are satisfied and, at the same time, the analysis criterion is maximised. Let us use $\mathbf{T}_x$ ($n \times m_1$) and T_y ($n \times m_2$) for rescaled **X** and **Y**-variables and use $\mathbf{t}_i$ for one transformed variable (either from **X** or from **Y**; $i=1,\ldots,m_1+m_2$), in addition use $\mathbf{C}_i$ for the set of transformations that satisfies the measurement restrictions of the i-th variable. Redundancy analysis with optimal scaling or REDUNDALS is

$$\text{minimise SSQ } (\mathbf{T}_x - \mathbf{T}_y \mathbf{V}\mathbf{W}')$$
$$\text{with } \mathbf{V}'\mathbf{T}'_y\mathbf{T}_y\mathbf{V} = n\mathbf{I},\ \mathbf{u}'\mathbf{t}_i = 0,\ \mathbf{t}_i'\mathbf{t}_i = n \qquad (5)$$
$$\text{and } \mathbf{t}_i \in \mathbf{C}_i \text{ for } i = 1, \ldots, m_1 + m_2$$

The transformation of a variable measured on a nominal level corresponds with the restrictions "objects with similar raw scores are similarly transformed". For a variable measured on an ordinal level the nominal restrictions must be satisfied but also the order restrictions, i.e. "the order of the transformed scores corresponds with the order of the raw scores" (secondary approach to ties, Kruskal and Shepard 1974). For variables measured on a numerical level all linear transformations of the raw scores satisfy the restrictions. For a more detailed description compare Young, De Leeuw and Takane (1976) or Van der Burg and De Leeuw (1983). As the number of measurement restrictions depends on the number of different raw scores or categories for each variable, the computer program which performs the REDUNDALS-analysis expects a small number of categories per variable. Therefore data with many different scores have to be recoded into a small number of categories. This is also done with the apple data described in the next section.

7.4 Apple Data

The data measured on apples have been collected by the ATO-DLO (Institute for Agrotechnology) (Koppenaal, 1991). During the season 1989-1990 objective measurement methods for determining the texture of apples were investigated. The question was whether instrumental measurement methods could be found or developed to predict the sensory qualities mealiness and firmness. Koppenaal (1991) used multiple regression to find the best model for predicting mealiness or firmness. Two varieties of apples have been used, namely Elstar and Cox's Orange Pippin. These apples have been grown and stored under different conditions, i.e. origin (low or high calcium), picking-date (early, middle, late), size (small, large) and storage temperature (3°, 13° or 23° Celsius). Because of the different conditions the apples will show differences in ripening and texture (see Table 1).

Table 1 Variables measured on two varieties of apples. For sensory variables the number of categories used for recoding in the analysis with ordinal restrictions is given. For instrumental qualities the minimum and the maximum score that occurred in Cox apples is shown together with the number of categories used for the recoding.

Background variables			Instrumental variables		min	max	cat
Ca	origin (low, high Calcium)		Pr	penetrometer: red side	2.5	5.8	4
Per	picking-date (early, middle, late)		Pg	penetrometer: green side	3.5	5.5	4
Si	size (small, large)		Pm	penetrometer: mean	3.7	5.9	4
Tmp	storage temp (3°, 13°, 23°)		Mo	expelled moisture	5.51	43.06	6
			Dr	dry matter	12.25	17.08	5
Sensory variables		cat	Ac	total titratable acid	3.55	8.07	5
M1	mealiness judge 1	4	It	Instron: thickness at failure	1.33	3.05	6
M2	mealiness judge 2	4	If	Instron: force at failure	26.21	71.67	5
M3	mealiness judge 3	4	Iu	Instron: area	11.51	56.64	5
F1	firmness judge 1	4	Is	Instron: slope	7.48	97.95	6
F2	firmness judge 2	4	Im	Instron: modulus	1.27	3.66	5
F3	firmness judge 3	4	Ct	catalase activity	6.90	19.40	6

The instrumental qualities investigated by Koppenaal (1991) are a firmness-measure as acquired by the penetrometer both at the red side and the green side of the apple, the amount of expelled moisture at compression, the amount of total titratable acidity, the amount of dry matter, the catalase activity and so-called Instron-measurements (thickness, force, area, slope, and modulus) (see Table 1). Instron is a universal testing machine to measure mechanical properties of food which can provide force/deformation plots. All these variables have been recoded into a small number of categories. The order of the raw scores has been maintained. However the categories were not equidistant. Especially raw scores with low occurrences were taken together.

The sensory qualities are the qualities firmness and mealiness as judged by three trained assessors on a rating scale varying from 0=soft/mealy to 100=firm/not mealy. These rating scales are recoded into 4 categories (1 (0-25), 2 (25-50), 3 (51-75), 4 (76-100)) (see Table 1).

The first analysis performed was a REDUNDALS analysis on sensory and instrumental qualities with ordinal restrictions. The second analysis was an analysis with numerical restrictions. For this analysis a different recoding has been used. The raw data were divided into many (ten to fourteen) equidistant categories.

7.5 Results For Cox Apples

Koppenaal (1991) who first analysed the apple data, used average z-scores of mealiness and firmness over assessors and performed two separate multiple regressions. In this study all individual sensory qualities are used in the criterion set. The predictor set consists of the instrumental measurements. In Table 2 the multiple regression coefficients are given resulting from a rank-2 REDUNDALS analysis of Cox apples with ordinal and numerical restrictions. In addition the regression weights (i.e. **VW′**) corresponding with the ordinal solution, are given for each variable. From Table 2 we see that all variables are predicted rather well. The difference between ordinal and numcrical is not very large. Further down we compare the solutions in more detail. The weights in Table 2 show that it is mainly the expelled moisture which predicts the sensory qualities. The next important predictor is Instron-slope. This result agrees with that of Koppenaal (1991). Because weights always control for the effect of the other predictors, we prefer to investigate correlations. As an orthogonal basis in the predictor space has been formed by means of **YV**, we are interested in the projections of the **X**- and **Y**-variables in this reduced rank space. These projections correspond with the correlations $n^{-1}\mathbf{T}'_x\mathbf{T}_y\mathbf{V}$ and $n^{-1}\mathbf{T}'_y\mathbf{T}_y\mathbf{V}$ (Figure 1a). In addition, we are interested in the projections of the apples into this space (see Figure 2a).

Table 2 Multiple correlation coefficients (MCC) for the Cox apple data and regression weights for the ordinal solution of a rank-2 REDUNDALS analysis.

	Mealy1	Mealy2	Mealy3	Firm1	Firm2	Firm3
MCC						
ordinal	0.628	0.724	0.604	0.736	0.539	0.610
numerical	0.573	0.728	0.632	0.594	0.558	0.650
Weights						
Pr	-0.057	0.106	0.145	-0.024	0.056	0.220
Pg	0.224	0.137	0.084	0.224	0.144	0.009
Pm	0.038	-0.200	-0.250	-0.014	-0.122	-0.352
Mo	-0.552	-0.547	-0.457	-0.600	-0.494	-0.354
Dr	-0.065	0.124	0.170	-0.027	0.066	0.257
Ac	-0.388	-0.222	-0.128	-0.384	-0.240	0.009
It	-0.130	-0.071	-0.039	-0.128	-0.079	0.008
If	-0.177	0.260	0.370	-0.093	0.128	0.576
Iu	0.055	-0.248	-0.312	-0.010	-0.149	-0.443
Is	-0.386	-0.381	-0.317	-0.419	-0.344	-0.244
Im	0.385	0.029	-0.099	0.337	0.112	-0.311
Ct	-0.032	-0.150	-0.167	-0.062	-0.107	-0.208

From Figure 1a we see that the sensory qualities have rather large correlations in the reduced rank space. We find that the sensory qualities are more grouped per assessor than per quality. However all six quality measures seem rather similar (vectors close to each other and of the same size). The instrumental qualities are mainly opposite. Expelled moisture is in the centre and has the longest vector. Therefore it is the best candidate for prediction. Pr and Pg seem to perform rather poor in this representation, however Pm, which is the mean of Pr and Pg, does well. The Instron-measures Im, Iu, Is and If can also have a real contribution to the prediction. Also Ac seems relevant. It, Dr, Pr, Pg and Ct are the worst candidates for prediction. The good predictors give a lot of redundant information (many predictors in the same (opposite) direction of the criterion variables). Most of it can be predicted by Mo and Is (see weights in Table 2).

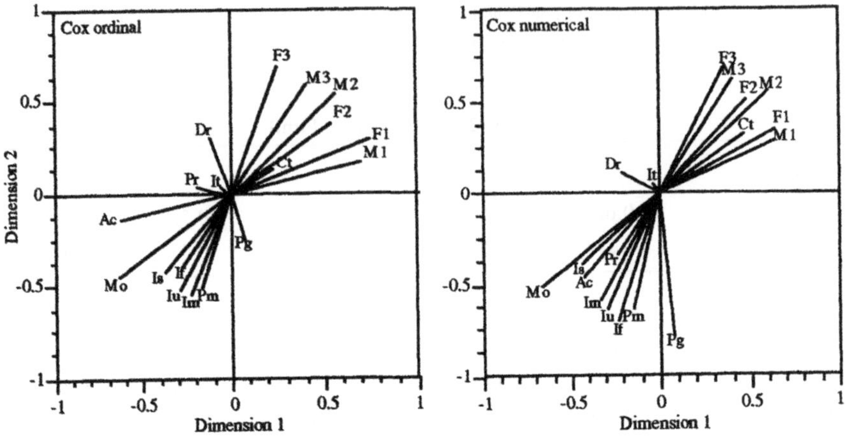

Figure 1 Correlations of the variables with the axes of the reduced rank space of the ordinal solution (1a: left) and the numerical solution (1b: right) for the Cox data.

In Figure 2a the Cox apples are projected in the reduced rank space. They are labelled by categories of expelled moisture. The lower the moisture the greater the firmness and the lower the mealiness. The two Figures (1a and 2a) can be thought on top of each other. Then each vector gives the direction of the high (rescaled) scores for the corresponding variable and the opposite direction points to the low (rescaled) scores. This holds only for long vectors like Mo, but not for short vectors like It. The plot of **YV**-scores and $n^{-1}\mathbf{T}'_x\mathbf{T}_y\mathbf{V}$ (which is equal to W) together form a biplot of the matrix **X** (Ter Braak, 1990).

To compare the results of an ordinal analysis with the results of a more standard technique like that from Van den Wollenberg (1977), a solution with numerical options is used. In this case recodings were taken which hardly reduce the information in the data (many equidistant categories). The result is given in Table 2 and Figure 1b. We saw already from Table 2 that the multiple correlation coefficients are not much better in the numerical case than in the ordinal case. Figure 1b shows that the configuration in the numerical case is also rather similar to that of the ordinal case. The figure is even more concentrated around the M2-axis and the sensory qualities are very close to each other. Thus averaging will not raise a big loss. Pg and Ct do much better in Figure 1b than in Figure 1a. In the numerical case the information from the predictors is also very redundant for the criterion variables. From Figure 1b, Mo and Pg will be chosen as best predictors of mealiness and firmness. This is what Koppenaal (1991) found.

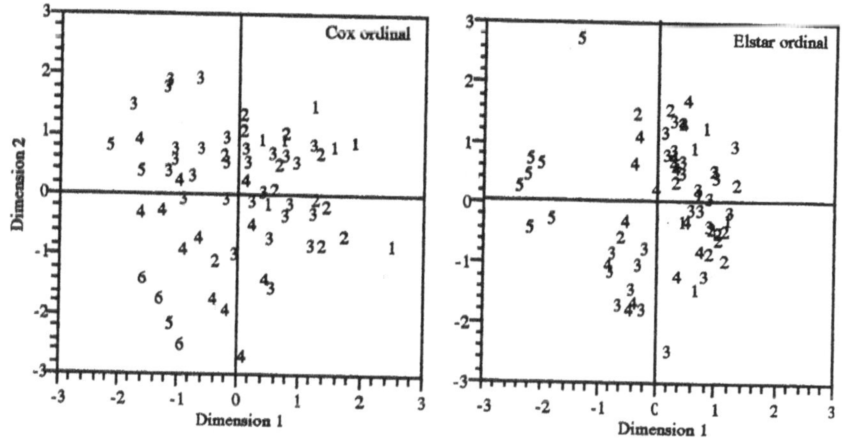

Figure 2 Projections of objects (apples) into the reduced rank space of the ordinal solution for the Cox data (2a: left) and the Elstar data (2b: right). The Cox apples are labelled by the categories of Expelled Moisture (1=low, ..., 6=high), the Elstar apples are labelled by titratable acidity (1=low, ..., 5=high).

A Procrustes rotation to match the correlations of Figures 1a and 1b showed a fit of 97.5% variance. So the two configurations are almost identical. Matching the configurations of the Cox apples in the reduced rank space for the ordinal and the numerical solution showed a fit of 86.1% variance. This is lower than the previous result of the correlations, but still an indication for strong similarity of the two solutions. It seems that the extra information used in the numerical analysis, compared to the ordinal analysis, is superfluous. This was found earlier by Van der Burg, De Leeuw and Dijksterhuis (1994) comparing an ordinal and a numerical solution of a generalised canonical correlation analysis of data from the sensory assessment of vegetable soups.

7.6 Results For Elstar

The Elstar data have been recoded in a way comparable with the Cox data. As the lowest and highest scores that occurred with Elstar apples differed from those of Cox apples, we used slightly different boundaries for the recodings into a small number of categories. Remarkable is that the variables of the Elstar data are much more skewed than the variables from the Cox data, especially many non-mealy apples are present.

For Elstar apples the same analyses were performed as for Cox apples. In Table 3 the multiple correlations as obtained from the REDUNDALS analysis are shown. From this table we see that the ordinal solution fits somewhat better than the numerical solution. The weights for the numerical solution (not given here) show that Pr, Ac, It, If, Iu, Is, Ct do better than the other variables. None of the weights is large, the maximum is 0.322. For the ordinal solution a similar pattern is found. However, in contrast with the numerical solution, Ac has low weights and Iu performs better than the rest. In this case the largest weight is 0.428.

Table 3 Multiple correlation coefficients for the Elstar apple data.

	Mealy1	Mealy2	Mealy3	Firm1	Firm2	Firm3
ordinal	0.685	0.706	0.615	0.509	0.821	0.497
numerical	0.453	0.545	0.507	0.393	0.633	0.485

Plots of the variables projected in the reduced rank space are given in Figures 3a and 3b. The Figures show a remarkable similarity (98.7% variance overlap with a Procrustes rotation). Both solutions show a strong one-dimensional structure. The criterion-vectors of the ordinal solution are a little bit longer than the ones of the numerical solution. This corresponds with the higher multiple correlations of Table 3. In the ordinal case the Mo-vector is placed is a little away from the other predictors. The different position of Mo may be due to the optimal scaling. As the predictor information is rather redundant, REDUNDALS tries to fit a second dimension by scaling the variable Mo as nonlinearly as possible. In this case a dichotomy results (1 versus 2 to 5). For the Elstar data it is reasonable to assume a rank-one restriction on the weights. We do not perform this analysis here. The result will correspond with the M2-direction of Figure 3a.

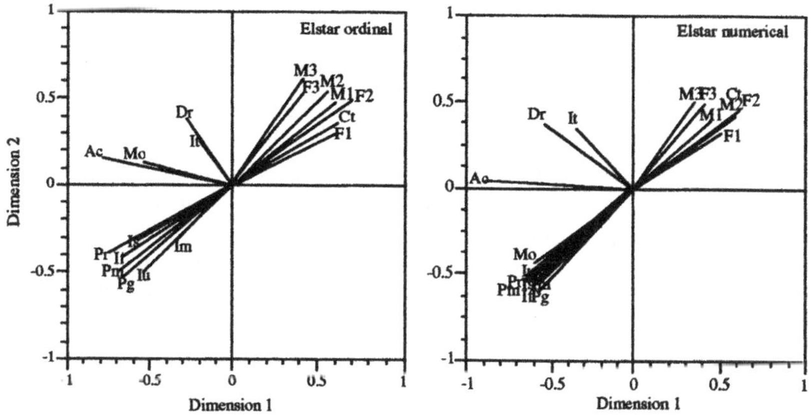

Figure 3 Correlations of the variables with the axes of the reduced rank space of the ordinal solution (3a: left) and the numerical solution (3b: right) for the Elstar data.

The plots of Elstar apples in the reduced rank space (Figure 2b ordinal solution, numerical solution not given) also show a high match (90.8% of the variance). The ordinal plot is more structured than the numerical plot. Figure 2b shows a cluster in the Ac/Mo-direction. Those apples are the only ones scored high on acidity. The apple highest on the second dimension has a unique pattern, scoring both high on acidity and dry matter. The clustering may be the reason that the ordinal solution fits better than the numerical solution. The fit is inflated by giving the objects with high scores for the original variables (here acidity and dry matter) an extreme position.

Good predictors for the sensory qualities are all variables projected in the M2-direction with long vectors (Pg, Pr, Pm, It, Iu, Ct). Thus we find Instron-measures and penetrometer-measures as Koppenaal (1991) did. Ac and Mo, which were good predictors of the Cox apples, do not satisfy for Elstar. Koppenaal too found that acidity is a good predictor for Cox apples and not for Elstar apples.

7.7 Conclusion

Using several multiple regressions Koppenaal (1991) concluded that penetrometer- and Instron-measures and expelled moisture are good predictors for Elstar and Cox apples. We obtained comparable results. He also found that Elstar did better than Cox (higher explained variance). However we find that Cox apples did better than Elstar apples: in the numerical REDUNDALS solutions Cox apples have higher multiple correlation coefficients than Elstar apples. Koppenaal did use only mealiness (average z-scores) in his regression analyses. His argument to drop firmness was that mealiness and firmness were practically interchangeable. We used both mealiness and firmness (without averaging). However, we do not think that this difference in approach explains why we find higher multiple correlations coefficients for Cox than for Elstar, as the mealiness- and firmness-variables turn out to be rather similar in the REDUNDALS-results too. Maybe the optimal scaling is responsible for this difference.

This study shows that nonlinear redundancy analysis can be a useful tool in analysing sensory-instrumental correlations. Especially if there are more than one criterion variables, redundancy analysis has the advantage of handling sets of variables instead of single variables like multiple regression, thus avoiding many separate analyses. The nonlinear options of the REDUNDALS technique prove useful in checking nonlinear relations between the variables. In this study it turned out that (at least for Cox apples) there was not much deviation from linearity.

CHAPTER 8

An Application Of Nonlinear Redundancy Analysis and Canonical Correlation Analysis

8.1 Introduction

Nonlinear redundancy analysis and nonlinear canonical correlation analysis are applied to a data set describing measurements on apples. The data were collected in order to find instrumental measures which predict the mealiness and the firmness of the apples. This study is a secondary analysis of the data. Results show that moisture, titratable acidity, penetrometer- and so-called instron-measures do well in predicting the quality of the apples. Two varieties of apples were used of which one does better in the analysis than the other.

By Eeke van der Burg and Garmt Dijksterhuis. Originally published in 1993 in *Psychometric Methodology*. Proceedings of the European meeting of the Psychometric Society, Trier, Germany. (R. Steyer, K.F. Wender, K.F. Widaman, Eds.), 74-79. (Reprinted with the kind permission from Gustav Fisher Verlag)

8.2 Techniques

Nonlinear redundancy analysis is a technique that predicts a set of criterion variables from a second set of variables. This prediction is done through maximising the explained variance of the criterion variables by a weighted sum of predictors, while at the same time all variables are optimally scaled. The rank of the weight matrix for the predictors is kept low (van der Burg and De Leeuw, 1990). Nonlinear redundancy analysis is a generalisation of the redundancy analysis of van den Wollenberg (1977). The computer program that performs nonlinear redundancy analysis is called REDUNDALS[1] (van der Burg and De Leeuw, 1990).

Nonlinear canonical correlation analysis is a technique that optimally relates two sets of variables with each other also using optimal scaling (van der Burg and De Leeuw, 1983). The corresponding computer program is called CANALS. Nonlinear canonical correlation analysis is a generalisation of ordinary canonical correlation analysis (e.g. Tatsuoka, 1988, Chap. 7).

Both REDUNDALS and CANALS are computer programs that fit in the system of nonlinear multivariate analysis as designed by Gifi (1990).

---

[1]An experimental program only.

8.3 Description of the Data

The data under investigation were collected to find instrumental measures which predict sensory qualities of two varieties of apples: Cox and Elstar (Koppenaal, 1990). The sensory qualities were measured by three trained assessors scoring "mealiness" and "firmness" on a line-scale ranging from zero to 100 (see Table 1). To obtain apples which differ in quality Koppenaal (1990) manipulated a number of variables which are referred to as background variables. The instrumental measures correspond with all sorts of push and pull measures applied to the apples by different machines (Instron, penetrometer). The scores have been recoded either into a small number of (ordered) categories, or into many equidistant categories. Both recodings were used, one to perform analyses with ordinal measurement restrictions and the second one (many equidistant categories) to perform analyses with numerical measurement levels.

Table 1 Variables Measured on two Varieties of Apples. For Instrumental Measures the Minimum and the Maximum Score that Occurred in Cox Apples is Shown.

Background variables		categories	
Ca	origin	low, high Calcium	
Per	picking-date	early, middle, late	
Si	size	small, large	
Temp	storage temperature	3°, 13°, 23° Celsius	
Instrumental variables		**minimum**	**maximum**
Pr	penetrometer: red side	2.5	5.8
Pg	penetrometer: green side	3.5	5.5
Pm	penetrometer: mean	3.7	5.9
Mo	expelled moisture	5.51	43.06
Dr	dry matter	12.25	7.08
Ac	total titratable acidity	3.55	8.07
It	Instron: thickness at failure	1.33	3.05
If	Instron: force at failure	26.21	71.67
Iu	Instron: area	11.51	56.64
Is	Instron: slope	7.48	97.95
Im	Instron: modulus	1.27	3.66
Ct	catalase activity	6.90	19.40
Sensory variables		**minimum**	**maximum**
M1	mealiness judge 1	0 (not mealy)	100 (very mealy)
M2	mealiness judge 2	0 (not mealy)	100 (very mealy)
M3	mealiness judge 3	0 (not mealy)	100 (very mealy)
F1	firmness judge 1	0 (firm)	100 (soft)
F2	firmness judge 2	0 (firm)	100 (soft)
F3	firmness judge 3	0 (firm)	100 (soft)

Table 2 Multiple Correlation Coefficients (MCC) for Cox Apples and an Indication for the Regression Weights of the Ordinal REDUNDALS Solution.

MCC		M1	M2	M3	F1	F2	F3
Ordinal solution		.63	.72	.60	.74	.54	.61
Numerical solution		.57	.73	.63	.59	.56	.65
Weights							
Pr	penetrometer: red side						
Pg	penetrometer: green side						
Pm	penetrometer: mean						•
Mo	expelled moisture	x	x	x	x	x	•
Dr	dry matter						
Ac	total titratable acidity	•			•		
It	Instron: thickness at failure						
If	Instron: force at failure			•			x
Iu	Instron: area			•			x
Is	Instron: slope	•	•	•	x	•	
Im	Instron: modulus	•			•		•
Ct	catalase activity						

x |weight| >0.4
• 0.3< |weight| <0.4

8.4 REDUNDALS Results

For the prediction of the sensory qualities the program REDUNDALS has been used, both with ordinal and numerical measurement restrictions. In Table 2 the results (multiple correlation coefficients (MCC) and the weights) are shown for the Cox apples. We see that Moisture and several Instron-measures have (relatively) high weights. The MCC's vary between 0.54 and 0.73, which is not bad. Ordinal and numerical MCC's do not differ very much from each other.

As weights in prediction techniques are influenced by the effects of the other variables within the same set, another way to examine REDUNDALS results is by making a plot of the reduced rank space. This is a subspace of low dimensionality spanned by the predictor variables. Both objects (in this case apples) and variables (predictors and criterion variables) can be projected into the reduced rank space. Figure 1 shows the results for Cox apples (ordinal REDUNDALS solution). Only the (optimally scaled) variables have been projected. We see from this figure that not only Mo and Is are important in predicting mealiness (M1, M2, M3) and firmness (F1, F2, F3), but that other variables are also good (Ac, Iu, Im, Pm).

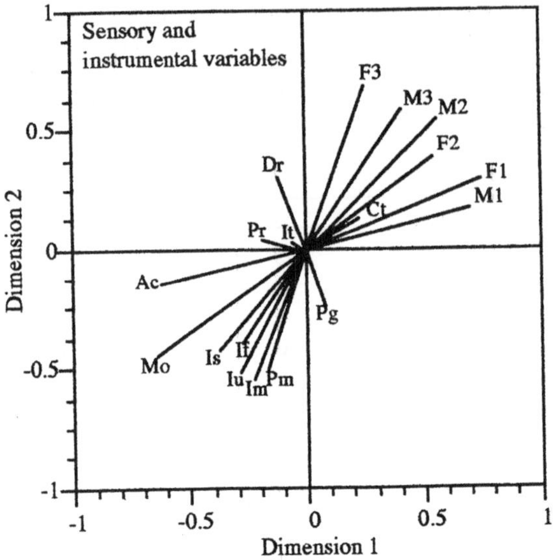

Figure 1 Projections of the optimally scaled variables into the reduced rank space for Cox apples (ordinal REDUNDALS solution).

The numerical REDUNDALS solution shows a lot of resemblance with the ordinal solution. The main difference is that the bundle of vectors is somewhat tighter, so that the solution is more one-dimensional. The fact that the difference between the ordinal solution with a few categories per variable and the numerical solution with many categories per variable is very small, shows that a lot of information in the data can be considered to be superfluous (in the sense that the technique does not use the information) (compare van der Burg, De Leeuw and Dijksterhuis 1994). Another reason to perform a numerical analysis is to keep results comparable with results from standard techniques. In this case with the results obtained by a program as suggested by van den Wollenberg (1977).

Elstar apples give results (not shown here) rather similar to those of the Cox apples. The ordinal solution is somewhat stronger than the numerical solution (higher MCC's). This difference is mainly due to the effect of the optimal scaling on objects with a unique score pattern. Moisture is not so important for Elstar apples as it is for Cox apples.

8.5 CANALS Results

Three analyses were performed with CANALS. The background variables were related to the sensory variables and to the instrumental measures. In addition, sensory and instrumental variables were compared. These CANALS analyses were performed mainly to check the underlying relations. Only the Cox apples were used for these analyses and, in addition, only ordinal measurement levels were considered.

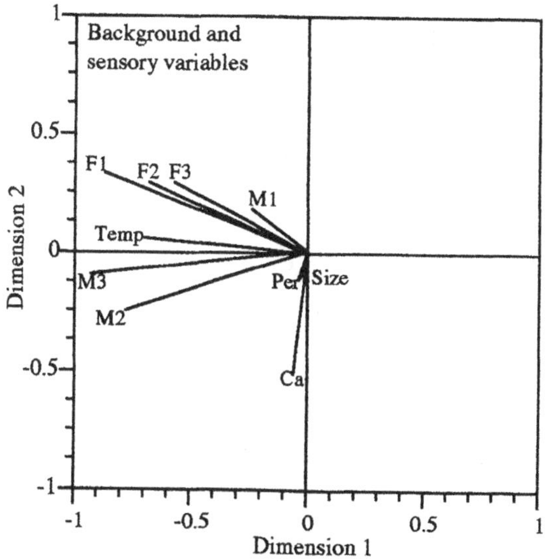

Figure 2 Projections of the optimally scaled variables into the canonical space of the sensory qualities (ordinal CANALS solution).

Analysis of the background variables and the sensory qualities show canonical correlations of 0.706 and 0.534. Projections of the variables into the canonical space of the sensory variables show that storage temperature (Temp) and the amount of Ca are mainly "responsible" for the different qualities of the apples (Figure 2). Background variables and instrumental measures show a much higher correspondence than background and sensory variables. In this case the canonical correlations are 0.927 and 0.869. From the projections of the variables into the canonical space of the instrumental measures (Figure 3) it can be seen that again Temp and Ca are the two background variables mostly related to the other set. Picking-date (Per) and size (Si) are not important. The third analysis (instrumental and sensory variables) was mainly a repetition of the redundancy analysis. The canonical correlations were 0.941 and 0.889, which seems really high. The projections into the canonical space show a picture which is comparable to Figure 1, therefore this plot is not given.

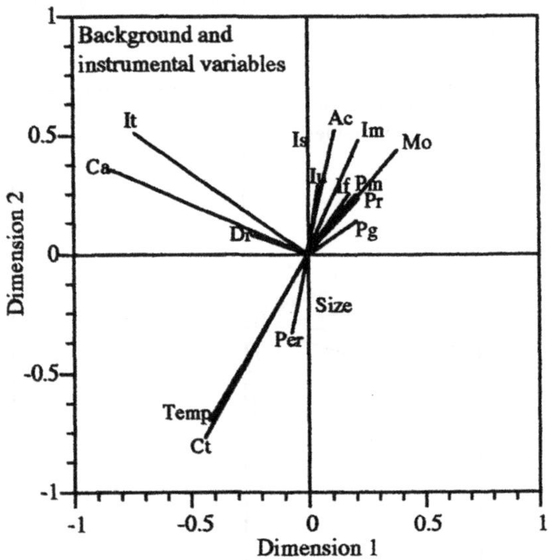

Figure 3 Projections of the optimally scaled variables into the canonical space of the instrumental measures (ordinal CANALS solution).

8.6 Conclusions

The prediction of the sensory qualities from the instrumental measures shows that moisture, titratable acidity, penetrometer and the Instron-measures are mainly related to mealiness and firmness. This result is comparable with that of Koppenaal (1990), who used the average over mealiness scores and the average over firmness scores in several multiple regressions.

This study shows that nonlinear redundancy analysis and canonical correlation analysis can be useful tools in analysing sensory-instrumental correlations. Especially if there are more criterion variables, redundancy analysis has the advantage of handling sets of variables instead of single variables like in multiple regression, thus avoiding many separate analyses. The nonlinear options of the REDUNDALS technique prove useful in checking nonlinear relations between the variables. In this study it turned out that (at least for Cox apples) there was not much deviation from linearity.

CHAPTER 9

Procrustes Analysis in Studying Sensory-Instrumental Relations

9.1 Introduction

The study of the relations between sensory judgements and instrumental, i.e. chemical or physical, measurements is a field where different kinds of Multivariate Analysis techniques can, and have been, applied. When the sensory data set contains only one variable, e.g. "acceptance", Multiple Regression can be used to relate the instrumental measures to this variable. In this situation PLS1 or Principal Component Regression are alternative techniques. In the general case each data set consists of a number of variables measured on a number of objects, say N, in this case foodstuffs. The two data sets are denoted here by $\mathbf{X}_1$ and $\mathbf{X}_2$ of respective orders $(N \times M_1)$ and $(N \times M_2)$, $\mathbf{X}_1$ contains sensory scores on M_1 attributes, $\mathbf{X}_2$ instrumental measures on M_2 variables. Another method often used to relate two data sets is PLS2. For the PLS techniques see e.g. Geladi (1988). A computer program based on the PLS philosophy is the Unscrambler (Tyssø *et al.* 1987, CAMO 1992).

When in a Multiple Regression situation both sets have more than one variable, Multiple Multivariate Regression, or Canonical Correlation Analysis results. When one of the sets is restricted in rank, a method called Redundancy Analysis appears (see e.g. van den Wollenberg 1977). For both Canonical Correlation Analysis and Redundancy Analysis non-linear versions are developed (van der Burg 1988). Linear, in this case means that the techniques assume underlying linear, interval or ratio-type scales. Linear techniques can be, and often are, used to analyse data of ordinal or nominal measurement level, but this may give unsatisfactory results. Non-linear methods analyse ordinal or nominal data by transforming the data in accordance with the measurement level. These transformations can be applied to one of the sets (semi-non linear, see Liu 1990) or to both sets (van der Burg and de Leeuw 1983, 1990). These methods are applied to sensory data in a number of cases (van der Burg and Dijksterhuis 1989, 1991, Liu 1990).

The above mentioned methods can be classified into two types: Asymmetric methods and Symmetric methods. The symmetry concerns the way the two data sets are treated by the method. *Asymmetric* methods try to predict one set from the other, and so treat both sets differently. PLS, Principal Component Regression, Redundancy Analysis and Multiple Regression are among these methods. *Symmetric* methods treat both sets identically, swapping the two sets makes no difference. Neither of the sets is tried to be predicted from the other, only the relations between the sets are studied. Examples of these methods are Canonical Correlation Analysis and Procrustes Analysis.

In this paper, Procrustes Analysis (Schönemann 1966, 1968, Gower 1975) is used to study sensory-instrumental relations. This method is more restricted than Canonical Correlation Analysis and Redundancy Analysis in that the only transformations allowed to the data sets are rigid-body transformations, i.e. transformation that respect the distances between objects. A linear version is used, though a non-linear version does exist (van Buuren and Dijksterhuis 1988) but not for all transformations. Generalised Procrustes Analysis is a well known method in the analysis of sensory data (e.g. Oreskovich *et al.* 1991), but it has not very often been used to study Sensory-Instrumental relations. Other methods, especially Canonical Correlation Analysis, sometimes give results with virtually perfect fit. This is partly due to the optimisation criterion of Canonical Correlation Analysis and with non-linear versions of the method partly to the freedom there exists in choosing non-linear transformations. The transformations of such techniques seem to be too lax, this is why in this chapter the much more strict transformations of linear Generalised Procrustes Analysis are applied to relate a sensory to an instrumental data set.

9.2 Data

The data consist of two data sets measured on 72 apples and is originally collected by the ATO Agrotechnology Institute in Wageningen, the Netherlands (see also van der Burg and Dijksterhuis 1993a). The apples were artificially manipulated to be different. The data set used in this paper consists of sensory judgements of three judges, judging the apples on Mealiness and Firmness only, giving in total 6 sensory variables. The instrumental variables are 7 different physical and chemical properties of the apples. The two sets have different variables, which is the same situation as with Free Choice Profiling data (Williams and Langron 1984; Arnold and Williams 1985).

The set of sensory data consists of 72 rows representing the objects, in this case 72 different apples of the race "Elstar". The 6 columns of this set are "Mealiness" and "Firmness" scores given to the apples by 3 assessors. One possibility of treating this data set is to average the variables over the judges resulting in only two variables, this was not done however. Averaging is only appropriate when the assessors use the variables in the same way, and from previous studies it is known that this assumption can often not be met (e.g. Dijksterhuis and Punter, 1990).

Usually each set contains the variables of one assessor and Generalised Procrustes Analysis is used to correct for the different use of the variables. In this paper the variables of the three assessors are put together in one set, and are treated as 6 different variables. The set of instrumental variables consists of 7 measurements of physical and chemical properties of the apples. Both the sensory and instrumental variables are shown in Table 1 together with some statistics.

Table 1 Mean, Standard Deviation, Range, Minimum and Maximum of the variables in the sensory and instrumental data sets. (PRED, PGREEN: penetrometer at red, resp. green side of the apple; MOIST: expelled moisture; DRYMAT: amount of dry matter; ACID: acidity; ITHICK; INSTRON-thickness; KATAC: catalase activity)

Variable	Mean	Std. Dev	Range	Minimum	Maximum
Sensory					
Mealiness 1	34.4	30.1	94	0	94
Mealiness 2	32.0	26.2	90	0	90
Mealiness 3	39.1	29.3	93	2	95
Firmness 1	51.4	21.6	88	6	94
Firmness 2	48.6	30.8	97	1	98
Firmness 3	52.2	27.5	91	5	96
Instrumental					
PRED	4.43	0.981	4	3	7
PGREEN	4.41	0.919	3.8	2.8	6.6
MOIST	24.5	13.1	39.65	10.18	49.83
DRYMAT	15.4	0.989	4.96	13.26	18.22
ACID	7.60	1.16	5.42	4.77	10.19
ITHICK	1.66	0.279	1.25	1.23	2.480
KATAC	12.1	2.17	9.9	7.2	17.1

In Table 1 can be seen that the sensory variables have more or less the same range, the instrumental variables however are very different. This is perfectly normal because they are measured with very different devices. It makes some sort of pre-scaling of these variables necessary for Procrustes Analysis (see e.g. Dijksterhuis and Gower 1991/2).

9.3 Procrustes Analysis

Procrustes Analysis was originally developed to match two different solutions from Factor Analyses (Hurley and Cattell 1962). *Generalised* Procrustes Analysis (Gower 1975) matches more than two data sets and is often applied to the data from different assessors from a sensory panel (e.g. Arnold and Williams 1985). For the study of sensory-instrumental relations the original Procrustes Analysis for two sets is used, but because the generalisation is a proper one, *Generalised* Procrustes Analysis with two-sets is identical to two-set

Procrustes Analysis. The differences between rotating to a target, or rotating both sets to a common set are of no concern in this study (see Schönemann 1966, 1968), and the abbreviation GPA will be used although strictly spoken the *G* for *Generalised* could be omitted. In this special case of two sets the Procrustes procedure is very simple and can be solved in one step, when there are more than two sets an iterative algorithm must be used.

The idea behind GPA is that the rows of each set define a configuration of the N object-points in multidimensional space, with the scores on the variables as coordinates on just as many dimensions. In this case there are $N=72$ points in $M_1=6$ dimensional space in the sensory set and 72 points in $M_2=7$ dimensions in the instrumental set. The distances between these object-points are seen as reflecting the similarity or dissimilarity between the objects represented by these points. In matching the sets $\mathbf{X}_1$ and $\mathbf{X}_2$ these distances are kept unchanged by the transformations applied to the configurations of points. The transformations applied by GPA are Translations, Rotations and Isotropic Scaling. The translations are chosen such that the centre of the configuration coincides with the origin of the multidimensional space.

9.3.1 Rotations

This transformation rotates the two configurations to maximum agreement. Maximum agreement is defined in terms of a least-squares criterion which measures the squared distances between corresponding points in the two configurations. For mathematical details on the method see Gower (1975), ten Berge (1977). It is important to realise that these rotations take place in the highest dimensional space possible, i.e. in our case in $M=\max\{M_1,M_2\}$. To make the two sets of the same order ($N \times M$) the smallest set is padded with zero columns. The result of the rotational procedure are two rotated configurations $\mathbf{X}_1\mathbf{H}_1$ and $\mathbf{X}_2\mathbf{H}_2$ with the squared distances between the corresponding points as small as possible. The average configuration $\frac{1}{2}(\mathbf{X}_1\mathbf{H}_1+\mathbf{X}_2\mathbf{H}_2)$, often called the *consensus* configuration or *Group Average*, exists in M dimensional space. In order to be able to study this configuration a Principal Component Analysis is applied to project the data onto a low dimensional space which can be plotted.

9.3.2 *Isotropic Scaling*

The isotropic scaling factors stretch and shrink the configuration to increase the fit further. They do not change relative distances between the corresponding points in the two configurations. Details about these scaling factors can be found in Gower (1975), ten Berge (1977) and Peay (1988). The resulting group average, with inclusion of the scaling factors ρ, looks like $\frac{1}{2}(\rho_1 \mathbf{X}_1\mathbf{H}_1 + \rho_2\mathbf{X}_2\mathbf{H}_2)$. The scaling factors correct for size differences of the configurations. Size being defined as total sum of squares of a configuration. When one set has a sum of squares different from the other, the scaling factors will shrink the bigger set relative to the smaller. This situation occurs in practice in sensory data analysis when the data sets stem from different assessors with differing scaling behaviour. In the example data set (see Table 1) it can be seen that the judges do not differ very much in their scoring behaviour, compared with the instrumental data set. The latter set is much more heterogeneous, the variables have different variances and the overall size of the set will depend mainly on the variables with the largest sums of squares. An isotropic scaling factor for the total set is not a good solution in this case, a scaling factor per variable would perhaps be better (see e.g. Næs and Kowalski 1989). However, such non-isotropic scaling factors complicate the mathematics of the procedure considerably (Commandeur, 1991).

9.4 A First Look at the Data: PCA

For a first look at the two data sets Principal Component Analysis is applied. Table 2 shows the percentage Variance Accounted For, for the analyses on the sensory and the instrumental sets separately.

Table 2 Percentage Variance Accounted For, for the Principal Component Analyses on the sensory and the instrumental sets separately.

Dimension:	1	2	3	4	5	6	7
Sensory							
%VAF	72.811	14.13	6.77	3.768	1.538	0.983	-
Instrumental							
%VAF	55.883	19.286	10.756	6.834	4.51	1.707	1.024

The table shows that for both data sets the first dimension explains a high percentage of variance, the second dimension explains much less. Higher dimensions can be inspected, but for ease of interpretation only the 2 dimensional results are used here.

9.4.1 *Sensory Variables*

From the numbers in Table 2 we conclude that there is a two dimensional space in which the data fit quite well (87% variance explained). Whether this space has a Mealiness and a Firmness dimension cannot be seen from this table. To see this the component loadings must be inspected. An easy way to do this is by means of plotting them in the two dimensional space (see Figure 1). The 72 apples' object scores are superimposed onto the same figure to see the relation between the apples and the variables.

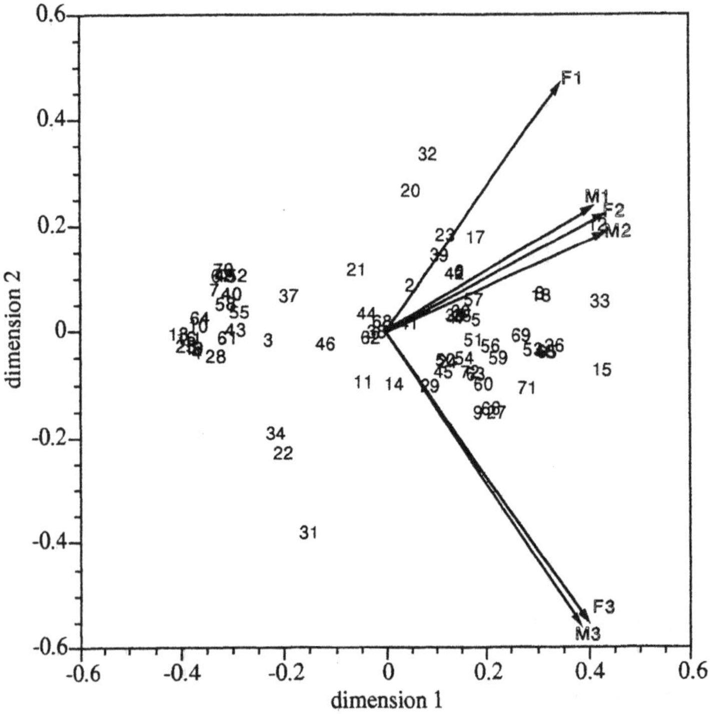

Figure 1 Sensory variables and the positions of the 72 apples after PCA of the sensory data set.

From Figure 1 can be seen that assessor 3 deviated from assessor 1 and 2. The variables Mealiness 3 and Firmness 3 (M3, F3) are used identically by this assessor, he or she uses them as if they were the same variables. Assessor 2 also uses Mealiness and Firmness alike (M2, F2), though he/she judges the apples different from assessor 3. The correlation between the Mealiness and Firmness variables of the same assessor is very large (see Table 3). Assessor 1 is the only one who uses the two variables differently (M1, F1), but they still have a correlation of 0.614.

Table 3 Correlations between the sensory variables.

	M1	M2	M3	F1	F2	F3
M1	1.000					
M2	0.798	1.000				
M3	0.614	0.645	1.000			
F1	0.640	0.648	0.429	1.000		
F2	0.827	0.917	0.625	0.668	1.000	
F3	0.579	0.638	0.926	0.444	0.645	1.000

The position of the variables suggest that the group of apples at the left-hand side of the centre of the figure are judged mainly not firm and not mealy. At the right-hand side lay the apples that have high Firmness and Mealiness scores.

9.4.2 *Instrumental Variables*

Table 2 suggests a good approximation of the instrumental data set in two dimensions too. A third dimension could perhaps be helpful, but for sake of simplicity we will only look at the two dimensional result (see Figure 2).

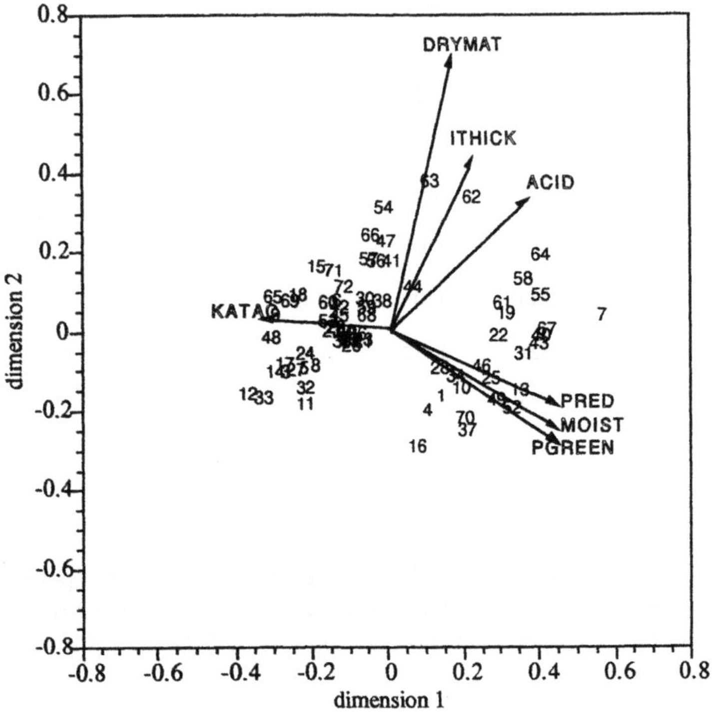

Figure 2 Instrumental variables and the positions of the 72 apples after PCA on the instrumental data set.

Again the apples' positions are superimposed onto the figure. In Figure 2 can be seen that there are two main directions: one defined by Drymat, Ithick and Acid, the other by Pred, Pgreen, Moist, and, negatively correlated, Katac. Two groups of apples can be identified, at the left the apples with high Katac scores and low Pred, Pgreen and Moist scores. At the right the apples with high Pred, Pgreen and Moist scores and low Katac scores. It looks as if the variables Drymat, Ithick and Acid are not very important in distinguishing these two groups. The correlations between the variables are presented in Table 4.

Table 4 Correlations between the instrumental variables.

	PRED	PGREEN	MOIST	DRYMAT	ACID	ITHICK	KATAC
PRED	1.000						
PGREEN	0.921	1.000					
MOIST	0.841	0.846	1.000				
DRYMAT	0.196	0.098	0.027	1.000			
ACID	0.609	0.526	0.502	0.534	1.000		
ITHICK	0.224	0.182	0.361	0.327	0.390	1.000	
KATAC	-0.609	-0.587	-0.669	-0.207	-0.497	-0.301	1.000

Now we have seen the structure of both sets separately, we can look whether GPA is able to relate the two sets to each other without sacrificing too much of this structure.

9.5 Matching the Sensory and Instrumental Data Sets

The sensory variables are scored on a line-scale ranging from 0 to 100. They do not need differential scaling because they have comparable scales. Their mean is set to 0, this is the translation operation from GPA which is the same as column centering the variables in this set. The instrumental variables are different and need differential scaling to make them comparable within the set of instrumental data. An obvious way of pre-scaling is converting the variables into z-scores. When the instrumental set is standardised in this way, there remains a size difference between the two sets. One could imagine this 'size' as the size of the cloud of object-points in the high dimensional space. When only the instrumental data set is converted into z-scores the sensory data set is much larger than the instrumental set. A simple solution is to standardise both sets prior the GPA. This leaves a size-(sum of squares) difference caused by the number of variables in the set. In order to correct for this, so-called P_k-scaling was proposed by Dijksterhuis and Gower (1991/2), which in addition removes the need for fitting isotropic scaling factors. In this study we applied no P_k-scaling and fitted isotropic scaling factors, only to find them very close to 1 for both sets (resp. $\rho_1 = 1.041$ and $\rho_2 = 0.964$).

The first two dimensions of the group average configuration resulting from GPA turn out to explain 68.8% variance. Figure 3 shows the loadings of the sensory variables in the GPA Group Average space and the positions of the apples after this set is rotated to fit the instrumental set.

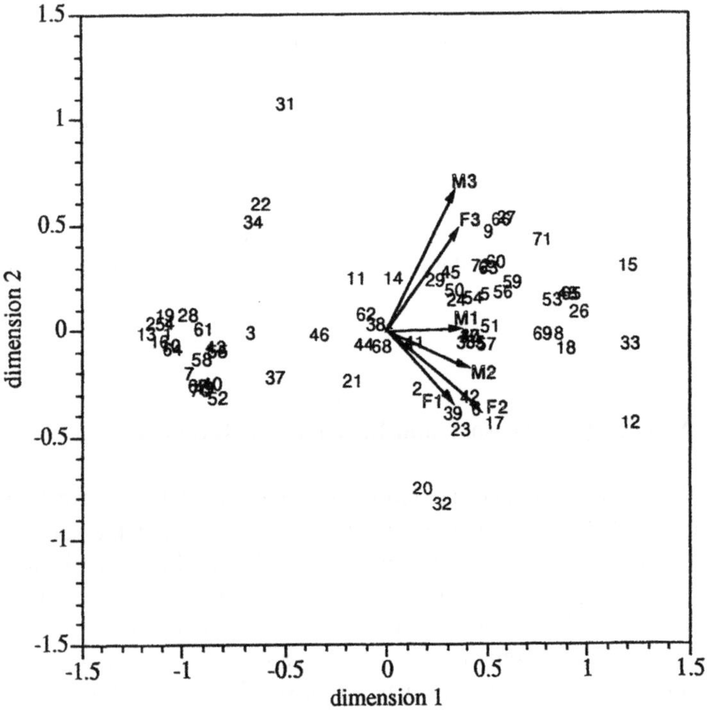

Figure 3 Sensory set with the variables and the apples after matching by GPA.

It can be seen from Figure 3 that the structure of the variables is like the structure revealed by PCA applied to the sensory sets separately (Figure 1). The two configurations have different orientations but these are irrelevant (one could use GPA to give them similar orientations, see e.g. Gower and Dijksterhuis 1994).

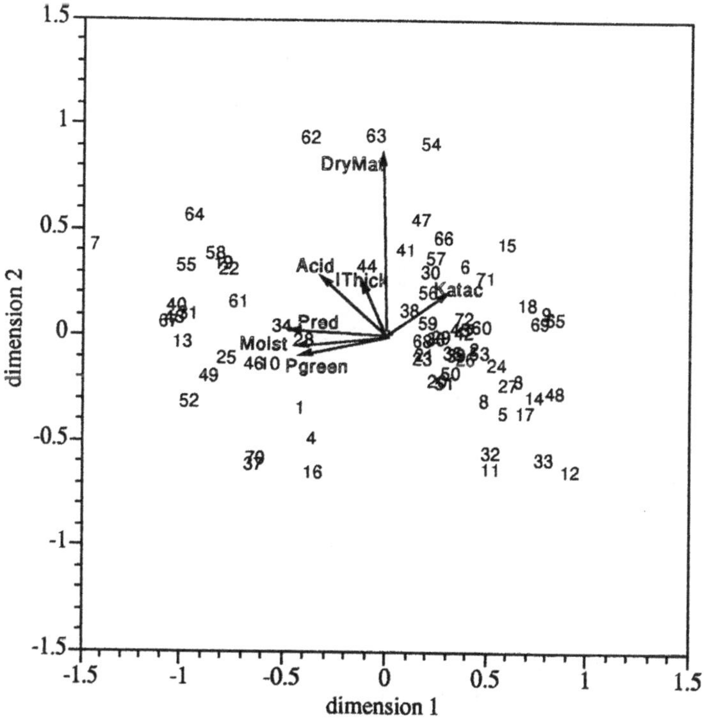

Figure 4 Instrumental set with the variables and the apples after matching by GPA.

Figure 4 shows the loadings of the instrumental variables and the apples after this set is rotated to fit the sensory set. Comparing Figure 4 with Figure 2 shows that, apart from orientation, the structure of the apple-configuration is not much different. For both the sensory and instrumental set the same two groups of apples can be identified.

Now GPA has matched the two sets to maximal agreement, the group average can be constructed by averaging the two configurations of apples. The variables remain in their position (see Figure 5).

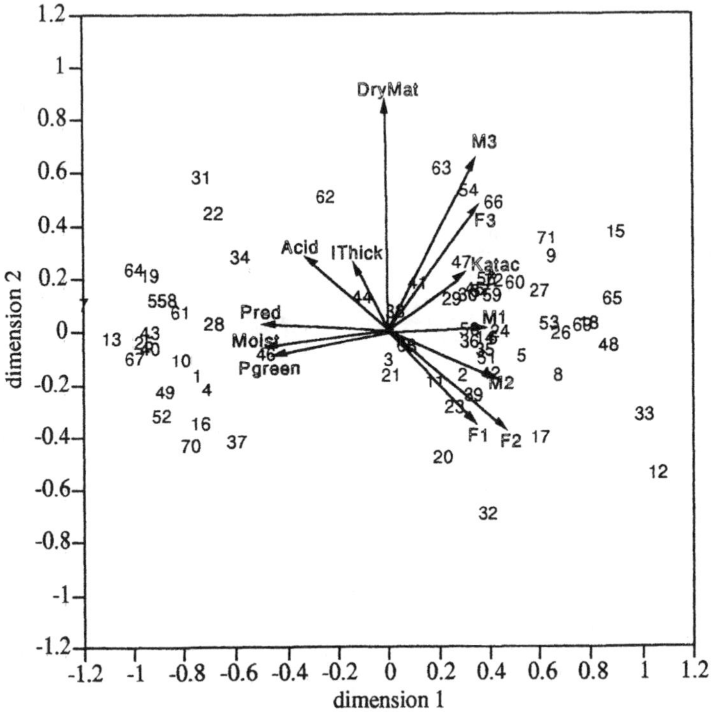

Figure 5 GPA group average with the positions of the apples and the sensory and instrumental variables.

The group average in Figure 5 shows the same two groups of apples and the position of both sensory and instrumental variables. It turns out that high Firmness and Mealiness scores coincide with high Katac scores. Low Firmness and Mealiness scores go together with high scores for Pgreen, Pred and Moist. The instrumental variables Acid and IThick have a less distinct relationship with Firmness and Mealiness. The variable DryMat shows no relation with the sensory variables.

9.6 Conclusion

Relations between a sensory and an instrumental data set can successfully be studied by means of a Procrustes Analysis. It turns out to be necessary to perform some kind of pre-scaling to the two data sets. In particular the instrumental variables need to be standardised or scaled because they can differ very much in range. The easiest way of scaling the two sets is to transform them into z-scores, so differences between variables within a set are removed. Other ways of pre-scaling can be useful too, e.g. P_k-scaling (see Dijksterhuis and Gower 1991/2), though not employed in this study. The structure of each set separately can be studied by PCA. GPA is able to match the sensory and instrumental sets while retaining most of the structure of the sets. Though GPA is not often found in sensory-instrumental studies, it turns out to be a useful tool to this end.

Concluding Remarks Part III: Sensory-Instrumental Relations

Summary

In this section the analyses in the three preceding chapters are summarised and compared. An overview of what part of the apple data is analysed using which analysis method is provided.

Introduction

Table 1 presents a brief overview of the different analyses applied to different parts of the apple data set.

Table 1 The different analyses performed in Part III (RA: Redundancy Analysis, CCA: Canonical Correlation Analysis, PCA: Principal Component Analysis, PA: Procrustes Analysis).

Chapter	Apples	Relations	Method	Measurement level	Figure
7	Cox	Sensory-Instrumental	RA	ordinal, numerical	1a/1b, 2a
7	Elstar	Sensory-Instrumental	RA	ordinal, numerical	3a/3b, 2b
8	Cox	Sensory-Instrumental	RA	ordinal	1
8	Cox	Background-Sensory	CCA	ordinal	2
8	Cox	Background-Instrumental	CCA	ordinal	3
8	Cox	Sensory-Instrumental	CCA	ordinal	not shown
9	Elstar	Sensory	PCA	numerical	1
9	Elstar	Instrumental	PCA	numerical	2
9	Elstar	Sensory-Instrumental	PA	numerical	3, 4, 5

Conclusions about the comparison of the different multivariate methods must be based on only partial comparisons for some methods.

Chapter 7
An Application of Nonlinear Redundancy Analysis

The main results of the Multiple Regressions (MLR) Koppenaal (1991) performed were replicated. One difference was that in the Redundancy Analysis (RA) the multiple correlation coefficients were higher for Cox apples than for Elstar, the MLR's resulted in a better result for Elstar than for Cox apples. New analyses could provide the reason for this small difference.

Redundancy Analysis appears to be a useful extension to Multiple Regression which was originally applied to the Apple data set. Two interesting properties of nonlinear RA in this respect are:

1. analysing more than one criterion variable, i.e. two data sets are analysed
2. inclusion of optimal scaling, so nominal, ordinal and numerical variables can be analysed.

Redundancy Analysis has an advantage over Canonical Correlation Analysis (CCA). With CCA (almost) perfect fit is obtained when two variables, each in one set, correlate very high. Despite the (almost) perfect fit the solution may explain a low amount of variance in the two data sets. The capitalisation on the correlation between -linear combinations of- variables in the two sets, renders CCA and Generalised Canonical Analysis (GCA) less fit for some two- or more-set problems with highly correlated variables in different sets. In Redundancy Analysis it is not a correlation coefficient which is maximised, but a linear combination of variables in the **X** data set is composed which explains as much variance as possible in the other, **Y**, data set.

Chapter 8
An Application of Nonlinear Redundancy Analysis and Canonical Correlation Analysis

The conclusion of the previous chapter is repeated here because the RA in this Chapter is the same as employed in Chapter 7. In addition CCA was applied, but with different data from the Redundancy analyses. CCA was used to study the relationships between the background variables and the sensory variables and between the background variables and the instrumental variables. It turns out that in both canonical correlation analyses the storage temperature and the Calcium level of the orchard the apples were grown, are important. These two variables seem to span a two dimensional predictor space.

The different use of RA and CCA reflect the differences between the two methods:

- RA tries to predict one data set from the other, here the sensory from the instrumental data
- CCA studies the relations -correlations- between the two data sets without predicting one from the other.

It can be argued to use RA to study the prediction of the instrumental and of the sensory variables from the background variables. Since the background variables are design variables, coded as dummies, CCA as applied is similar to a Multivariate Analysis of Variance with four factors (Storage temperature (three levels), Ca-level (two levels), Picking date (three levels) and Size (two levels)) and respectively six sensory or 12 instrumental variables.

Chapter 9
Procrustes Analysis in Studying Sensory-Instrumental Relations

In this chapter, Procrustes Analysis was applied to the Elstar apple data set. First both the Sensory and the Instrumental data set are inspected by means of Principal Component Analysis. Biplotting the results provides a useful view of the data. The Figures (Chapter 9, Figure 1 and 2) show the 72 apples and their position relative to the variables. Subsequent matching of the two data sets by means of Procrustes Analysis produces Figures 3 and 4. These two Figures are combined into the Group Average (Figure 5). The Group Average shows both the apples and the instrumental and sensory variables in one plot. The positions of the apples and variables in this plot can be interpreted and prove meaningful.

PCA turns out to be a useful method to provide an initial look at the individual data sets. The objects and variables can be (bi-)plotted and give a useful overview of the data. Outliers and other anomalies become visible in such plots, turning these plots into a handy device to check the data. When the individual plots are in order, they may be subsequently matched, in this paper by GPA, but other methods (RA, CCA, see the other chapters in this part) can be used too.

The transformations of GPA are such that the relative distances between the object-points (i.c. apples) remain intact. CCA and RA apply transformations which do not keep distances intact. GPA uses more restricted transformations than CCA and RA which will generally result in a lower fit. The advantage GPA may have over CCA and RA is that there is hardly any chance of finding an artificial perfect fit with real data.

Discussion

Table 1 shows that in Chapter 7 ordinal and numerical solutions of Redundancy Analysis are compared for the two races of apples. A high agreement of the two solutions emerged, as measured by a Procrustes rotation of the ordinal and the numerical solution towards each other. Table 2 summarises this fact. Both the configurations of the object scores -the apples- and of the component loadings -the variables- are rotated towards each other.

Table 2 Percentage of variance common to the ordinal and numerical solutions from the Redundancy Analysis applied in Chapter 7, as measured by a Procrustes Rotation.

source	Cox	Elstar
correlations	97.5	98.7
object scores	86.1	90.8

Table 2 shows a high agreement of the ordinal and numerical solution. It is to be expected that the agreement is high for both the correlation- and the object-score configuration, when there is a high fit. A low Procrustes fit would have meant different solutions, which would probably show in both the correlation- and the object-score configuration.

Conclusion

Nonlinear MVA is a useful method for the analysis of Sensory-Instrumental relations. However, too much freedom in fitting the data may result in an artificial fit, and this is why GPA may sometimes be a more appropriate and straightforward method. When the Sensory-Instrumental relations can reasonably be expected to be nonlinear, nonlinear methods clearly have an advantage over linear methods. In practice a balance between fitting nonlinear models, by means of optimal scaling, and fitting straightforward linear models (e.g. GPA) should be sought. Achieving this balance can be difficult and careful analyses must be carried out by researchers, keeping in mind not only the statistical methods but also the properties of the matter under research.

Suggestions for Future Research

Some lines of future research are indicated, both from a Sensory (-Instrumental) and from a statistical point of view.

Sensory

The study of Sensory-Instrumental relations could be continued in a number of directions of which two are:

1. Analysis of existing data material
2. Carefully collect new sensory and instrumental data

Many available Sensory-Instrumental data sets exist which have been collected by sound methods but have been underanalyzed, for example, only one aspect of the data has been examined, and/or only "classical" methods have been used. These data sets could be re-analysed using the methods used in this part ((non-)linear Generalised Canonical Analysis, Redundancy Analysis and Generalised Procrustes Analysis in Chapters 7, 8, 9) and other methods of (non-linear) MVA.

The second way could be employed when there is a particular sensory aspect of a certain product that needs investigating. The collection of future Sensory-Instrumental data sets should be done using an appropriate design, focusing on aspects thought to be important, and using already available knowledge about the products. Take for example a certain off-flavour in a beverage. A number of sound and defective samples of the beverage can be tasted by a panel, and chemical and physical measurements can be carried out. By repeatedly presenting the panel with a different set of (off-flavoured) beverage samples, and carrying out the instrumental measures, the off-flavour could be tracked to a particular cause, say a particular component due to oxidation, contamination or bacterial spoilage. The exploratory data analyses presented in this chapter, using the MVA methods Canonical Correlation Analysis, Redundancy Analysis and Generalised Procrustes Analysis lend themselves to such an approach.

Multivariate Data Analysis

The Biplot (Gabriel 1971, Gower 1992b) is a useful way to gain insight in the data. Biplots, briefly mentioned in Chapter 9, could be incorporated in a large number of MVA methods (see e.g. ter Braak 1990). The plotting of both the products and the sensory attributes as well as the instrumental variables, is a promising tool for the exploration of Sensory-Instrumental data. Biplots can be incorporated in MVA in different ways (see e.g. Chapter 4). Illustrating the use of Biplots in a Sensory-Instrumental data analysis context would be useful to forward this method and would give the researcher a useful device to inspect her/his data.

PART IV

TIME-INTENSITY DATA ANALYSIS

INTRODUCTION TO PART IV

Summary

The theme of the fourth and last part of this book is an interesting direction of research in Sensory Science, viz. Time-Intensity research.

In the first chapter (Ch. 10) non-centered Principal Component Analysis is proposed as an analysis method of a number of TI-curves collected of the same products. In Chapter 11 three PCA methods (on correlations, on covariances, on raw data) are compared for the analysis of TI-curves. The last chapter (Ch. 12) is a brief chapter suggesting a new direction in modelling TI-curves, focusing on the *shape* of the TI-curve.

Chapter 10
Principal Component Analysis of Time-Intensity Bitterness Curves[1]

The data analysed in this chapter are from an experiment in which K subjects record the perceived bitterness intensity by moving a slider on a monitor using a computer-mouse. Usually average TI curves are calculated to give a representation of a product-specific TI curve. The problem is that there often are large individual differences, so the average TI curve is probably not a good representation. An alternative approach is to perform a Principal Component Analysis on a matrix containing the individual TI-curves in its columns. The resulting, so-called, Principal Time Intensity Curves (*PTIC*'s) theoretically are better representations than the average curves. Sometimes the PTIC's for the different products are hard to distinguish. In such cases a variant PCA method, non-centred PCA, gives results which show more differences between the products.

There is a terminological flaw present in this chapter. In §10.5 there is mention of "the loadings in **Q**". The matrix **Q** however does not contain *loadings* but it contains the *weights*. The term *loadings* is usually reserved for the elements of the matrix $\mathbf{Q\Phi}$. In the application presented here there is not much difference in interpreting weights or loadings, the difference is in the scaling of the axes by the corresponding singular value.

---

1 Principal Component Analysis of the matrix with TI-curves was originally developed at OP&P's in cooperation with Dr. Stef van Buuren (van Buuren 1991, 1992). When the application of this method on a new TI-data set posed some problems in interpreting the results (Dijksterhuis 1991, Dijksterhuis and Krabbe 1991) an alternative PCA method was used. This was the non-centred PCA method of Chapter 10. The method was presented at the "2nd Agro Industrie & Methodes Statistiques Conference" held in Nantes, France in June 1991 (Dijksterhuis 1991), and later the paper in the Journal of Sensory Studies was written based on this presentation.

Chapter 11
Principal Component Analysis of TI-Curves: Three Methods Compared[2]

In the study reported here the Time-Intensity curves are analysed by means of three variants of Principal Component Analysis (PCA):

- PCA on correlations;
- PCA on covariances;
- PCA on raw data.

The three variants differ in the amount of *a priori* standardising applied to the curves. The resulting Principal Curves can be interpreted and reflect underlying similarities or differences between the time courses of the tastes involved. The loadings[3] from the PCA's can be used to help interpreting the Principal Curves and to identify clusters of assessors, and outliers. Non-centred PCA retains both level and variability information from the TI-curves, and may be the preferred method for this reason. The centred PCA variants seem to give the tightest clustering of Principal Curves, hence differences between the Principal Curves may not become visible.

---

[2] In 1992 an experiment was carried out at OP&P's by Margo Flipsen, a student from the department of Marketing and Marketing Research of the Agricultural University of Wageningen, the Netherlands. This research resulted in a report (Flipsen 1992), from which the paper in Chapter 11 resulted. Some of the results of the Flipsen research were presented by Pieter Punter as part of a more general talk on Time-Intensity research, at the "Understanding Flavour Quality" symposium (Punter 1992), in Bristol, UK. The paper in Chapter 11 is printed in Food Quality and Preference, for the "Understanding Flavour Quality" symposium issue (Dijksterhuis *et al.* 1994). Paul Krabbe deserves mention here for advise and help with the analyses. Els van den Broek is thanked for comments on earlier versions of this paper.

[3] The same terminological flaw reported in the previous paragraph for Chapter 10 is present in Chapter 11. What are referred to as loadings are commonly called weights.

Chapter 12
Matching the Shape of Time-Intensity Curves[4]

Time-Intensity Curves are often summarised by average curves, or recently by Principal Curves. In this chapter a different method is proposed in which the *shape* of the curves is the central concept. A method to match the individual TI-curves by stretching or shrinking the curves is suggested. The stretching and shrinking is performed by isotropic scaling factors which are computed for each curve. These factors can be tabled and inspected. When a large number of scaling factors are computed it will be hard to study them in a table. In such cases they can be represented in a plot, for which in this paper PCA is suggested. In this plot a structure in the scaling factors for the different TI-curves may become visible. It is also possible to make a plot in which the homogeneity of the group of assessors can be studied.[5]

---

[4] Originally the method described in Chapter 12 was presented at the "3rd Agro Industrie & Methodes Statistiques" Conference held in Montpellier, France in November/December 1992 (Dijksterhuis 1992). It was one of the methods in a joint project of OP&P and Unilever Research Laboratories, Vlaardingen, the Netherlands. The project was largely carried out by Els van den Broek, a student from the department of Marketing and Marketing Research of the Agricultural University of Wageningen. This project resulted in a report (van den Broek 1993) and two articles (van den Broek *et al.* 1994, Dijksterhuis and van den Broek 1995) of which this chapter contains the latter.
Dr. Arne Maas (Unilever Research Laboratorium, Vlaardingen, the Netherlands) is thanked for comments on an earlier version of this article.

[5] This chapter suffers the same terminological flaw as the previous two chapters, what are called loadings are usually called weights.

CHAPTER 10

Principal Component Analysis of Time-Intensity Bitterness Curves

Garmt Dijksterhuis

(Originally published in 1993 in *Journal of Sensory Studies*, 8, 317-328. (Reprinted with kind permission from Food & Nutrition Press)

CHAPTER 10

Principal Component Analysis of Time-Intensity Bitterness Curves

10.1 Introduction

The study of the change of taste over time, termed Time-Intensity is now well known (see e.g. Lee and Pangborn 1986, Overbosch *et al.* 1986, Neilson 1957). In this chapter the Time-Intensity data stem from an experiment in which 5 assessors score the perceived bitterness of drinks. Often the resulting TI-curves are averaged over the assessors and measures as e.g. Maximum Intensity value, Time to reach this value, Area under the curve, etc., are computed (see e.g. Punter *et al.*, 1989). In this chapter Principal Component Analysis (PCA) of the TI-curves (van Buuren 1991, 1992) is evaluated as an alternative to averaging the curves and an alternative PCA method which may in some cases provide extra information is suggested.

10.2 Data

The data consist of the intensity scores recorded by the TI-equipment each second, during 90 seconds. In this particular experiment 6 drinks were presented to the judges. The drinks were mixtures of two caffeine concentrations (B1, B2) and two sugar concentrations (C1, C2). The following solutions were presented to the subjects: B1, B2, B1C1, B1C2, B2C1, B2C2, see Table 1. B1C1 was presented twice to give a possibility for inferring some kind of replication validity of the data.

Table 1 Concentrations of the components of the six drinks used as stimuli.

bitter drinks:	.75 g/l caffeine	1.5 g/l caffeine
	B1	B2
bitter/sweet mixtures:		
C1 20 g/l sugar	B1C2	B2C1
C2 40 g/l sugar	B1C2	B2C2

A TI-curve for drink i (i=1,...,N) made by judge k (k=1,...,K) is represented here by a column vector x_{ki} which has length 90. It is a column of the 90 recorded intensity values. The data of K judges for one particular drink i are collected in a matrix X_i. Figure 1 schematically shows this matrix. It contains 5 of the above mentioned columns of 90 intensity values. There is such a matrix for each of the six stimuli.

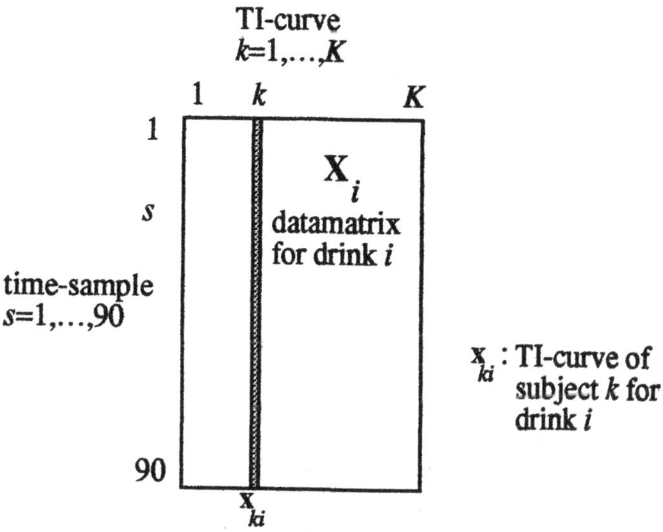

Figure 1 Datamatrix $\mathbf{X}_i$ with the column-vector x_{ki} containing the TI-curve of subject k for drink i.

In this case there were 90 time-samples, one each second. Of course the number of time-samples is not important to the analyses presented here. The objects of interest in this case are the 6 different drinks. The research question is how perceived bitterness changes differently over time for the different mixtures.

Figure 2 shows the 5 curves for one of the drinks (B2C2), each curve is an averaged curve of 4 presentations to a judge. Because TI-curves from the same individual are highly similar (e.g. see Overbosch *et al.* 1986, Figure 4, p. 335) it was assumed that it is permissible to average the curves from the 4 presentations of the same drink. These averaged curves are a slightly smoother version of the raw curves.

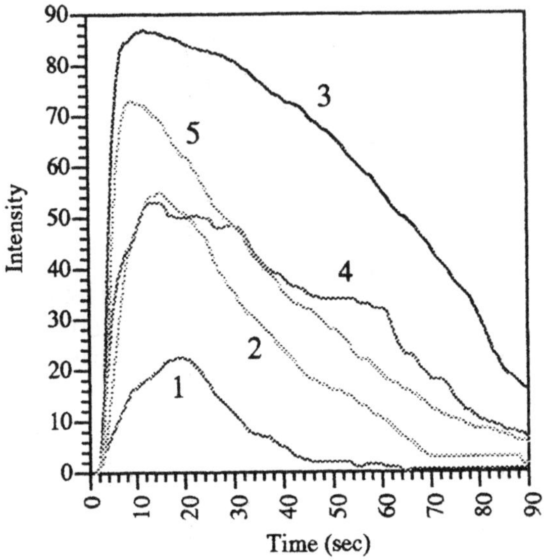

Figure 2 Individual TI-curves from the 5 judges for the same drink (B2C2).

Figure 2 shows that there are rather large differences between the curves from different individuals, although they stem from the same drink. The very idiosyncratic shape of a subject's curve is sometimes called the *signature* of the subject (van Buuren 1991). Another experiment with real drinks (not artificially made up from bitter/sweet mixtures) showed a more apparent signature effect (Dijksterhuis and Krabbe 1991). In Figure 2 the curves differ in overall height and in shape. Subject 3 shows a much slower decline than subject 5 and subject 4 seems to experience some kind of after-taste effect at 40 to 60 seconds

In Figure 3 the TI-curves from one subject for the 6 different drinks are presented.

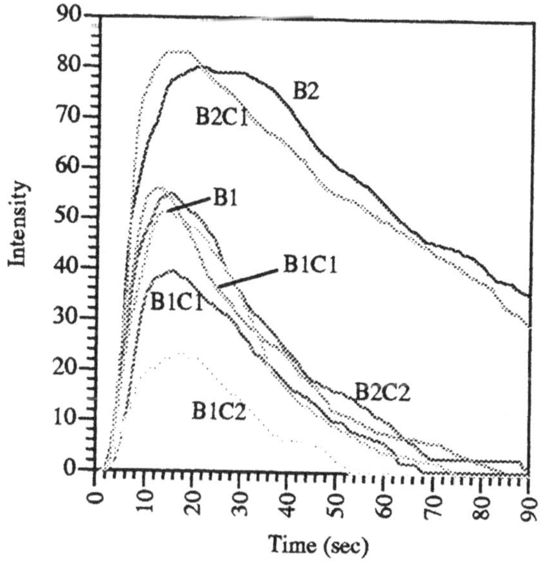

Figure 3 TI-curves from one judge for the different drinks.

From Figure 3 it is immediately clear that B2 and B2C1 are the most bitter drinks, B2C1 has an earlier and higher peak and a faster decline. Perhaps the sugar interacted with the bitter to result in this difference, but this is just the result for one subject. The next most bitter curves are the ones for B1, B1C1 and B2C2. The replicated B1C1-curves are not exactly the same, one has a much higher peak than the other. Clearly B1C2 is the lowest in perceived bitterness.

From Figures 2 and 3 it is apparent that all curves have the same overall shape, one of the most obvious differences seems to be the general level of the curve. Often the curves from the same subject are very much alike in shape (the aforementioned signature effect). Usually it is not the subject but the product that interests the researcher. The goal of many a TI experiment is to get information about the change over time of a particular attribute judged by the subject (bitterness in this case). To this end average product-curves are often calculated. In the notation of this paper the average curve for object i is $\bar{\mathbf{x}}_i$

$$\bar{\mathbf{x}}_i = K^{-1}\sum_{k=1}^{K} \mathbf{x}_{ki} \tag{1}$$

In this way average curves can be calculated for each drink i. Of course some information can be extracted from the average curves. Figure 4 shows the average curves for the drinks.

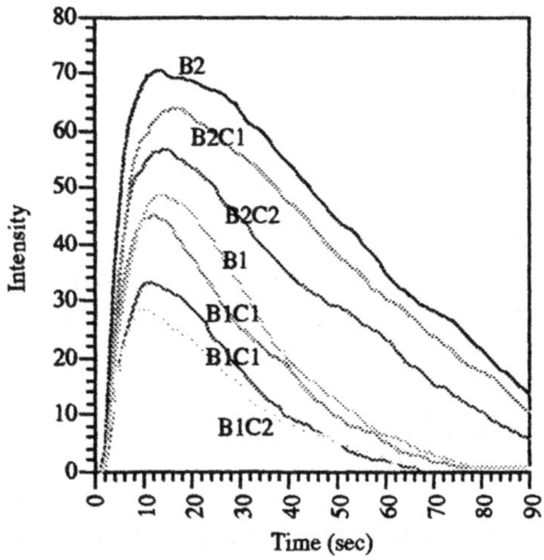

Figure 4 Average TI-curves for the different drinks.

In Figure 4 it can be seen that all curves with the most bitter component (B2) have a higher level than all B1 curves. Even the B2C2 (with the strongest sugar solution) lies higher than the B1 curve.

Realising that the individual curves have very different shapes, averaging over the individuals does not necessarily result in a very good representation of the curves. In addition some other fundamental problems arise when averaging TI-curves (see Liu and MacFie 1990, MacFie and Liu 1992).

10.3 Principal Curves

An average variable can be seen as an unweighted linear combination of a number of variables. Including weights in (1) gives

$$\bar{\mathbf{x}}_i = K^{-1}\mathbf{X}_i\mathbf{a} \tag{2}$$

which gives the average when **a** is a column-vector of length K consisting of only ones. A better representation would result from **a** containing weights that are chosen to maximise the variance in the "average" curve, which would in that case not be the "average" curve anymore but a weighted average curve.

Van Buuren (1991, 1992) suggests performing a Principal Component Analysis of the matrix $\mathbf{X}_i$ which contains the TI curves as columns. He performs a Singular Value Decomposition of the matrix after it is column-centered. The column centered matrix $\mathbf{X}_i$ is decomposed:

$$\mathbf{X}_i = \mathbf{P\Phi Q}' \tag{3}$$

with **P** and **Q** orthogonal matrices and Φ a diagonal matrix with singular values, decreasing in size down the diagonal. This decomposition describes what is known as a Principal Component Analysis on the Covariance matrix. It is equivalent to performing an Eigenvalue decomposition of the covariance matrix $\mathbf{X}'_i\mathbf{X}_i$.

Of course an analysis based on the correlation matrix could also be carried out. Then the data, i.e. the TI curves in $\mathbf{X}_i$, are not only centered but also divided by their standard deviations, they are transformed into z-values. With such $\mathbf{X}_i$ the matrix $\mathbf{R} = \mathbf{X}_i'\mathbf{X}_i$ contains the correlations between the curves in the matrix. Because the variance, and hence standard deviation, of the curves is believed to contain potential interesting information, the decomposition is not carried out on the correlation matrix. Later it will be shown that even the mean values contain information that had perhaps better not have been subtracted from the data.

The decomposition (3) gives a new set of curves $\mathbf{P\Phi}$ which are called *principal curves*. When an eigenvalue decomposition of the covariance matrix $\mathbf{X}_i'\mathbf{X}_i$ is carried out, like

$$\mathbf{X}_i'\mathbf{X}_i = \mathbf{Q\Phi}^2\mathbf{Q}' \tag{4}$$

the new set of curves can be found in $\mathbf{X}_i\mathbf{Q}$.

It can be shown easily that (3) and (4) are the same by substituting $\mathbf{X}_i = \mathbf{P\Phi Q}'$ into (4), which gives:

$$\mathbf{X}_i'\mathbf{X}_i = \mathbf{Q\Phi}'\mathbf{P}'\mathbf{P\Phi Q}' \tag{5}$$

which is $\mathbf{Q}\mathbf{\Phi}^2\mathbf{Q}'$ because $\mathbf{P}'\mathbf{P} = \mathbf{I}$ and $\mathbf{\Phi}' = \mathbf{\Phi}$ is diagonal. Because also $\mathbf{Q}'\mathbf{Q} = \mathbf{I}$, post-multiplying both sides of (3) with $\mathbf{Q}$ gives $\mathbf{X}_i\mathbf{Q} = \mathbf{P}\mathbf{\Phi}$.

Usually not all principal curves will be of interest, only the first two or three might be interpretable. These two or three principal curves will approximate the information in the matrix $\mathbf{X}_i$ to a certain extent, and it is very useful to know how good this approximation is. This is relatively easily seen from the diagonal matrix $\mathbf{\Phi}$ which contains the singular values $\sqrt{\Phi_j}$, or from $\mathbf{\Phi}^2$ with eigenvalues Φ_j, assumed to be ordered from large to small. The square root of the singular values is introduced here to simplify the notation in the remainder.

The amount of information of all curves in $\mathbf{X}_i$ represented by the first r principal curves, also called *fit*, is the sum of the first r eigenvalues:

$$\vartheta_r = \sum_{j=1}^{r} \phi_j \tag{6}$$

Equivalently, of course, the *loss* of this representation is

$$\phi_r = \sum_{j=r+1}^{R} \phi_j = tr\Phi^2 - \vartheta_r \tag{7}$$

with R the rank of the matrix (see e.g. Johnson 1963).

It is good practice to explain the fit of a representation in terms of the relative amount of variance. In this case it boils down to $(\vartheta_r/\mathrm{tr}\Phi^2)$, the *percentage* of Variance Accounted For (VAF) is of course $(\vartheta_r/\mathrm{tr}\Phi^2)100\%$.

The analysis of the covariance matrix focuses on the general shape of the curves for a drink. When the curves all have more or less the same shape this method comes up with one major principal curve with a relatively large ϑ_1 which has this shape. Van Buuren (1991, 1992) presents an analysis of 48 curves from different drinks and different subjects together. The resulting first three principal curves explained respectively 83.3%, 10.2% and 4.0% of the variance. Other analyses (Dijksterhuis and Krabbe 1991) resulted in comparable percentages.

Figure 5 shows the first three principal curves of the drinks presented in the previous figures and Table 1.

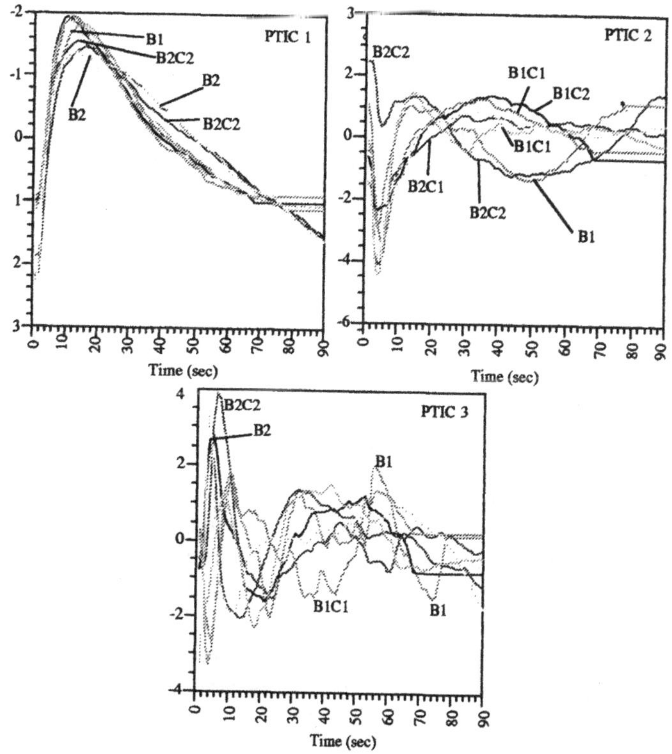

Figure 5 First, second and third principal curves for the different drinks.

In an attempt to use the principal curve method to compare the differences in perceived bitterness changing over time for different drinks Dijksterhuis and Krabbe (1991) ran into trouble using the first two or three principal curves obtained by analysing the matrices $\mathbf{X}_i$. The same problem is illustrated by Figure 5 in which it is not easy to distinguish between the curves of the different drinks. There seem to be roughly two kinds of curve in PTIC1, B2 and B2C2 against the other ones. In PTIC2 it is not easy to get to an interpretation that makes sense, PTIC3 seems to contain mostly noise.

Figure 3 showed that one of the major differences between the drinks lay in the general level of a curve. The average curves (Figure 4) also show an important level difference. Because the method of principal curves works on column-centered curve matrices $\mathbf{X}_i$ most of the level-information is lost entirely. Though the method of van Buuren (1992) is a welcome new tool in investigating the differences in TI-curves this aspect proves less fortunate. There is however, an alternative method which is closely related to the principal curve method.

10.4 Non-Centered PCA

An alternative method of analysing the matrices with TI-curves $\mathbf{X}_i$ was found in a method called non-centered PCA. As the name suggests the matrix $\mathbf{X}_i$ is analysed without it being put in deviations from the mean first. The decompositions are the same as in the previous paragraph, the only change is that $\mathbf{X}_i$ now is an uncentered, *raw* data matrix. Like in the column-centered case a new set of curves is found in $\mathbf{P\Phi}$. These vectors are no principal components in the usual sense of PCA and they will be called *non-centered principal curves*. When the mean contains substantial information, i.e. the mean value is relatively large, the first non-centered principal curve will reflect the position of the mean (see e.g. Jolliffe, 1986; van de Geer, 1986). This can be seen in Figure 6 where the first non-centered principal curves (NPTIC's) are very much like the mean curves in Figure 4.

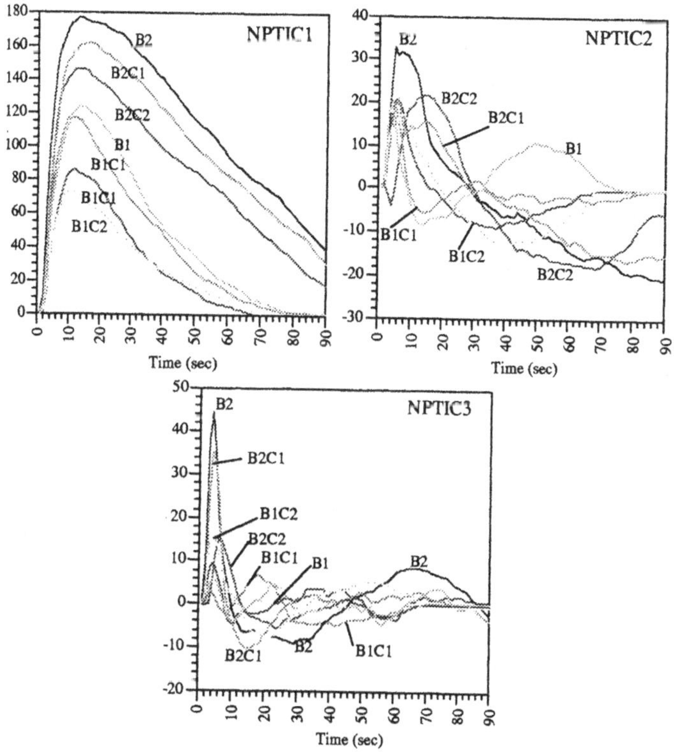

Figure 6 First three non-centered principal curves for the different drinks.

The second NPTIC shows B2 having the highest peak at about 10 seconds. B1 distinguishes itself clearly from the rest at 40 to 70 seconds, perhaps reflecting an after-taste phenomenon. It is strange that B2 does not show this, being a more bitter solution. B2 does show something of the kind in NPTIC3 at approximately 70 seconds. The mixtures with sugar and B2 seem to have a later peak in NPTIC2 (B2C1, B2C2). Perhaps sugar masks a very bitter taste for some 12 seconds? However wild the interpretations, the second and third non-centered principal curves are more apart than the second and third principal curves in Figure 5 and appear to contain less noise. Flipsen (1992) compared the alternative PCA analyses and concluded that the non-centered PCA shows the most information.

The expressions for the fit ϑ_r and loss σ_r of a representation with r curves are not different from the centered case (see (6) and (7)). There is a complication with the explained variance concept of an uncentered PCA. Variance is usually measured around a certain mean value, but with uncentered PCA the mean is not subtracted, it is still in the data. This makes is difficult to compare the different PCA approaches on ground of the amount of variance they explain. Other matters, such as the interpretability of the solution, must be taken into account to compare the methods.

10.5 Further Considerations

Another alternative to van Buuren's (1992) principal curve method is row-centering the $\mathbf{X}_i$ instead of column centering them. The row means are just the mean curves $\mathbf{x}_i$ while the column means contain an average intensity value of each curve for the K subjects.

The representation of the new curves $\mathbf{P\Phi}$ is one result of the analysis, the so called component scores. Another result are the loadings in $\mathbf{Q}$, each subject k has a loading on the principal curves. A plot of these loadings reveals differences between the subjects and can provide interesting information about such things as panel-homogeneity and outliers. This aspect is being studied and will be presented in future papers (Flipsen 1992, Dijksterhuis *et al.* 1994).

Still another possibility is performing a PCA (however centered) on the matrices $\mathbf{X}_k$ ($k=1,\ldots, K$), instead of $\mathbf{X}_i$ ($i=1,\ldots, N$), which contain the curves of one subject for the different drinks. In that case $\mathbf{P\Phi}$ contains principal curves for a subject, the so called signature. $\mathbf{Q}$ then contains the loadings of the drinks on these principal curves. Some unpublished research (Dijksterhuis, 1991) showed that these loadings, stemming from different subjects and after (Procrustes-) matching (Gower, 1975), can give interpretable results.

Since these kind of analyses of Time-Intensity data are very new, much research is needed to establish its usefulness.

CHAPTER 11

Principal Component Analysis of Time-Intensity Curves: Three Methods Compared

11.1 Introduction

Time-Intensity studies often result in a large quantity of data. A traditional way of aggregating the data is averaging the curves over the assessors. Subsequently a number of parameters can be computed from the averaged curves (see e.g. Yoshida 1986, Leach and Noble 1986, Punter *et al.* 1989). However, because individual differences are often quite large (see e.g. Overbosch *et al.* 1986, Pangborn *et al.* 1983, Schmitt *et al.* 1984) the average TI-curve is not a good representation of the individual TI-curves (see e.g. Liu and MacFie 1990, MacFie and Liu 1992). Overbosch *et al.* (1986) suggested a

---

By Garmt Dijksterhuis, Margo Flipsen and Pieter Punter. Originally published in 1994 in Food Quality and Preference, 5, 121-127. (Reprinted from Food Quality and Preference, 5, Dijksterhuis, G.B., Flipsen, M., Punter, P.H., Principal Component Analysis of Time-Intensity Curves: Three Methods Compared, 121-127, Copyright (1994), with kind permission from Elsevier Science Ltd, The Boulevard, Langford Lane, Kidlington OX5 1GB, UK) way of aggregating TI-curves based on a different treatment of the rising and the falling part of the curves. However Liu and MacFie (1990) noted some drawbacks of this method: it does not allow for a plateau of maximum intensity, not for more than one peak in the curve, nor for non-zero end-points, which are things that often occur in practice. Another disadvantage, noted by Van den Broek (1993) is the fact that the method only results in an aggregated curve and no reference to the individual curves seems to be provided.

Principal Component Analysis (PCA, see e.g. Jolliffe 1986) of TI-curves is a method in which the differences between individual curves are taken into account (van Buuren 1991, 1992; Dijksterhuis 1993). Resulting *Principal Curves* should be better representations of the individual curves than the average curve. Furthermore PCA has some additional properties which may prove useful in interpreting TI-curves.

11.2 Method

To investigate the usefulness of PCA for TI-data analysis, TI-data for four sweet solutions and four bitter solutions were collected (Flipsen 1992). All solutions were prepared in drinking-water and evaluated in fourfold at room temperature. The trained panel consisted of one male and eight female assessors. All participated in previous TI-experiments. A series of six training sessions preceded the actual measurements. The concentrations of the stimuli are shown in Table 1.

Table 1 Concentrations of the bitter (tetrahop and caffeine) and sweet (sugar and saccharin) stimuli.

	Concentration 1	Concentration 2
Tetrahop	12 mg/l	25 mg/l
Caffeine	125 mg/l	160 mg/l
Sugar	30 g/l	60 g/l
Saccharin	0.1 g/l	0.2 g/l

Concentrations are set at roughly the same sweetness ratio as the level of the sugar sweetness scale and at the same bitterness. Time Intensity was measured with the PSA-computer-system (OP&P 1991). The perceived intensity was recorded on a vertical line scale anchored at the bottom end by "weak" and at the upper end by "very strong". The assessors swallowed (part of) the sample

(40ml) and then immediately recorded the perceived intensity for 90 seconds. The samples were presented in eight sessions, four with sweet solutions and four with bitter solutions. Mineral water and crackers were provided for neutralising between samples. Each sample was presented four times to a subject in random order. For each assessor an average Time-Intensity curve over the four replications for each stimulus is calculated.

11.2.1 Average TI-curves

Figure 1 shows the average sweetness curves of the four sweet stimuli.

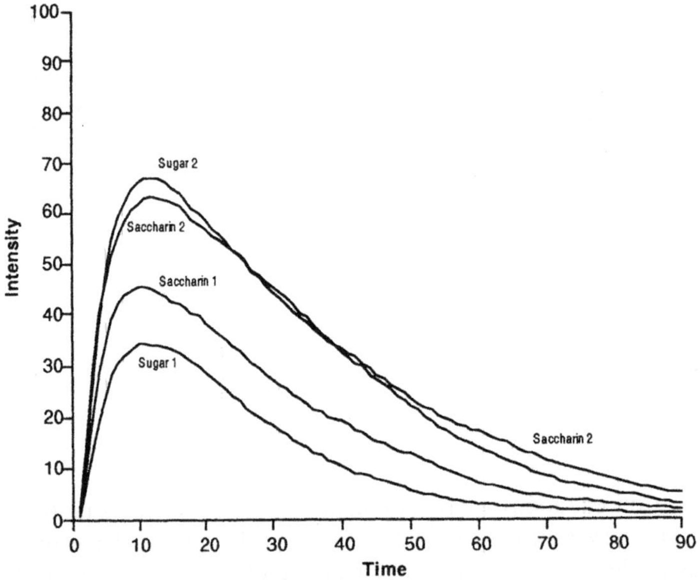

Figure 1 Average curves for the sweet stimuli.

The Sugar 2 concentration is perceived as the most intense sweet stimulus (at Imax). Saccharin 2 is perceived about as sweet as Sugar 2. Saccharin 1 is perceived more sweet than Sugar 1.

The average bitterness curves are presented in Figure 2.

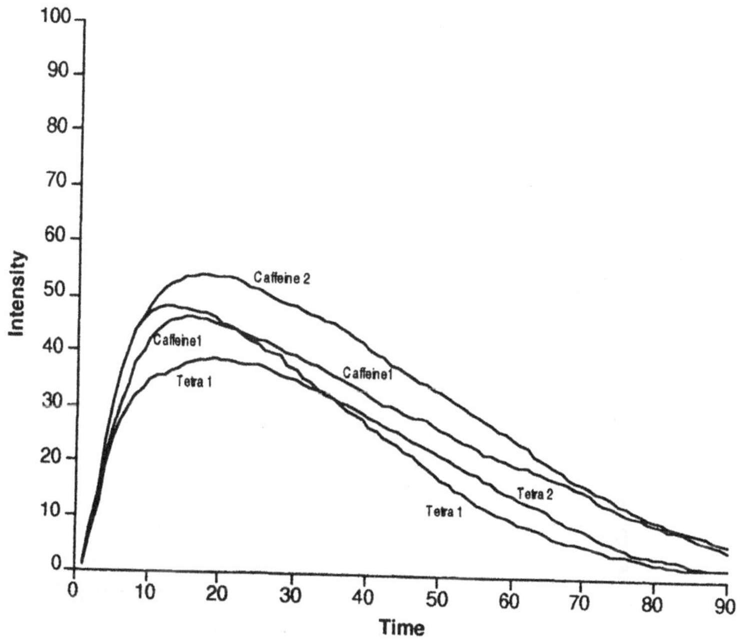

Figure 2 Average curves for the bitter stimuli.

The higher concentrations of all stimuli are perceived as more intense. The figures 1 and 2 show that the average curves of the sweet stimuli have a different shape from the average curves of the bitter stimuli. The sweetness curves have a faster rate of rising and a faster rate of decline than the bitterness curves. Average time to maximum intensity is 11-13 seconds for sweet and 17-18 seconds for bitter. Differences between sweet and bitter are perhaps related to the location of sensitivity for the four qualities of taste. Sweet is being sensed at the front part of the tongue and bitter at the back part.

11.2.2 Individual Differences

There are substantial differences between assessors for the same stimulus. Typically, each assessor demonstrates an idiosyncratic curve shape which is distinct and reproducible (see e.g. Overbosch *et al.* 1986). Usually the TI-curves differ both in overall height and in shape. Figure 3 shows the individual curves of the nine subjects for the high sugar concentration.

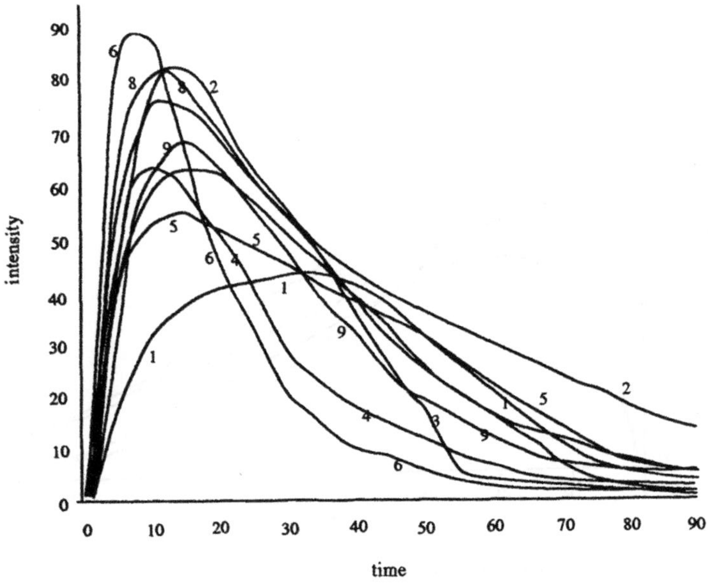

Figure 3 Individual subject's curves for the high sugar concentration (Sugar 2, numbers in the curves are subject-numbers).

The idiosyncratic shape of a subject's curve is called *signature* of a subject by Van Buuren (1992). A consequence of the high between-subjects variability is that the average curves drawn in Figure 1 and 2 are not necessarily a good representation of the individual curves.

11.3 Principal Curve Analysis

Principal Component Analysis is called *Principal Curve Analysis* in the context of TI-data analysis. Applying Principal Component Analysis to a matrix with TI-curves was originally suggested by van Buuren (1991). PCA weights individual curves in such a way that similar curves receive large weights, while deviating curves will receive low weights and will not substantially affect the resulting principal curve (Van Buuren 1992). The first principal curve is the weighted average of the individual curves that explains most of the total variation. The weights which the individual curves receive are called *curve loadings*. These loadings show how the curves contribute to each principal curve and may be used to classify assessors and interpret the obtained principal curves.

11.3.1 PCA Variants

Three variants of PCA are compared, the variants differ in the amount of initial transformations on the data. The most common transformations are:

- *centering*: The average level is subtracted from each individual TI-curve, the result is a TI-curve with positive and negative values around an average value of zero. A PCA on centered data analyses the covariances between the TI-curves.
- *standardising*: When in addition to centering the data are normalised too, this is commonly called standardising. The TI-curves are divided by their standard deviation and consequently have a variance of one. A PCA on centered data analyses the correlations between the TI-curves.

When none of the above is applied the raw data remain. A PCA on the raw, untransformed data is called a *non-centered PCA*. This method was suggested by Dijksterhuis (1993) to analyse TI-curves, the curves resulting from this analysis are called *Non-Centered Principal Curves*. Note that both level (average) and variability information is retained.

Van Buuren (1992) performed a PCA on centered TI-curves, consequently the covariances between the individual curves are analysed. This results in a new set of curves which will be called *Covariance Principal Curves*. The covariances contain information on the variability of the curves, the level is removed.

Usually PCA is performed on the correlations between variables, here between the individual TI-curves. In this case the TI-curves are standardised curves. The curves obtained by PCA are now called *Correlation Principal*

Curves. Both level and variability information is removed from the curves, what remains could be considered the underlying *shape* of the curves.

The heart of the PCA computations is the so called Eigenvalue Decomposition or the related Singular Value Decomposition, but these are beyond the scope of this paper. Van Buuren (1992) and Dijksterhuis (1993) provide some information on technical matters concerning PCA (see Jolliffe 1986 for more on PCA and for further references). The "goodness of fit" measure associated with Principal Component Analysis, usually presented as an Eigenvalue or a percentage of Variance Accounted For (VAF), can be used with Principal Curve Analysis too. This measure however depends on the centering of the data, so it is hard to use it to compare the uncentered with the centered results. The analyses reported here all had more than 90% VAF by the first two Principal Curves.

11.4 Non-Centered Principal Curves

In this section the Non-Centered Principal Curves of the sweet and bitter solutions are computed. In addition the loadings for the sweet solutions are displayed and used in interpreting the obtained principal curves.

11.4.1 Non-Centered PCA of TI Data From Four Sweet Solutions

In Figure 4 the first Non-Centered Principal Curve for each sweet solution is shown.

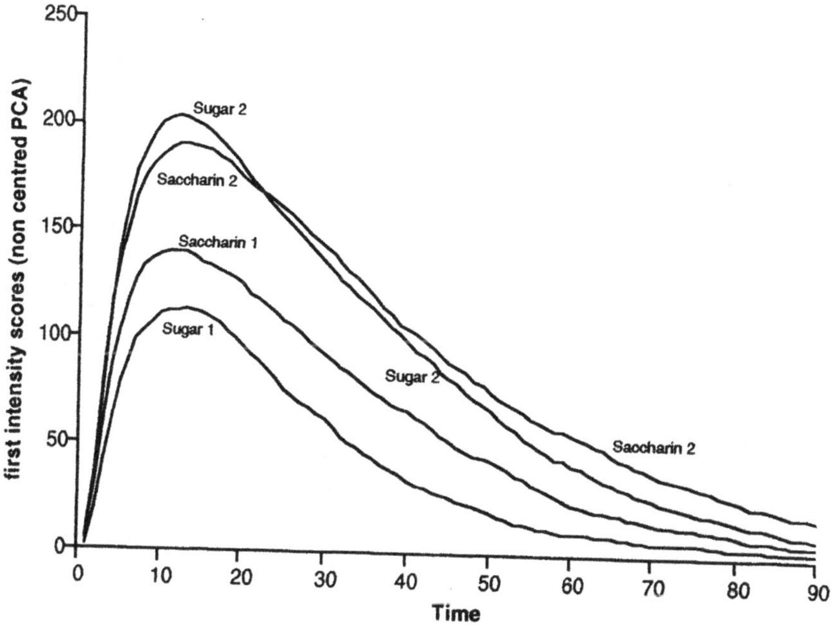

Figure 4 First Non-Centered Principal Curves for the sweet stimuli

The shape of the first Non-Centered Principal Curves resembles that of the average curves (compare with Figure 1 and 2). Apparently, the different weights for the individual curves hardly changed the shape of the average curve. In Figure 5 the second Non-Centered Principal Curves are displayed.

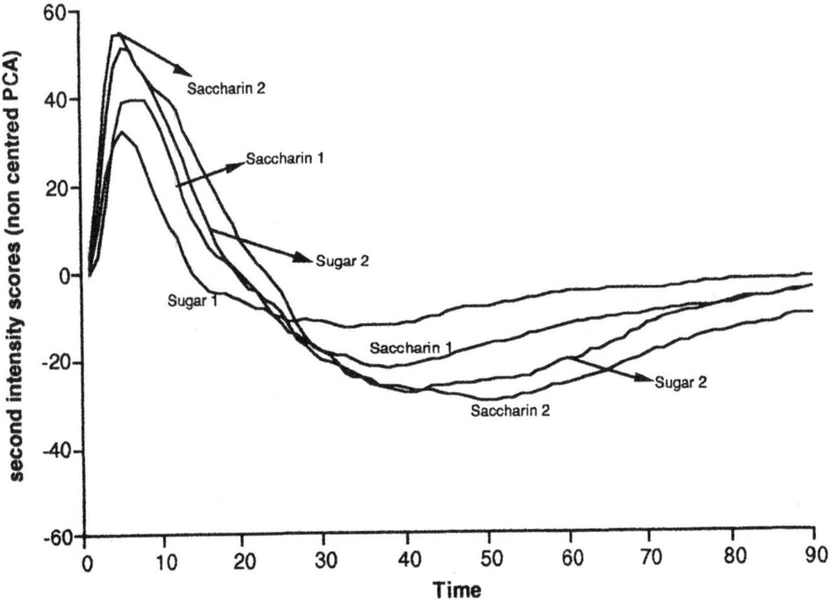

Figure 5 Second Non-Centered Principal Curves for the sweet stimuli.

The second Non-Centered Principal Curves have a maximum followed by a minimum. It turns out that the second Non-Centered Principal Curve models the rate of rising and declining of the individual TI-curves.
With the aid of the curve loadings on the first two Non-Centered Principal Curves and the individual TI-curves we can interpret the first two Non-Centered Principal Curves. In Figure 6 the curve loadings of the nine assessors are given for the sugar solutions. Each point represents the TI-curve of one assessor.

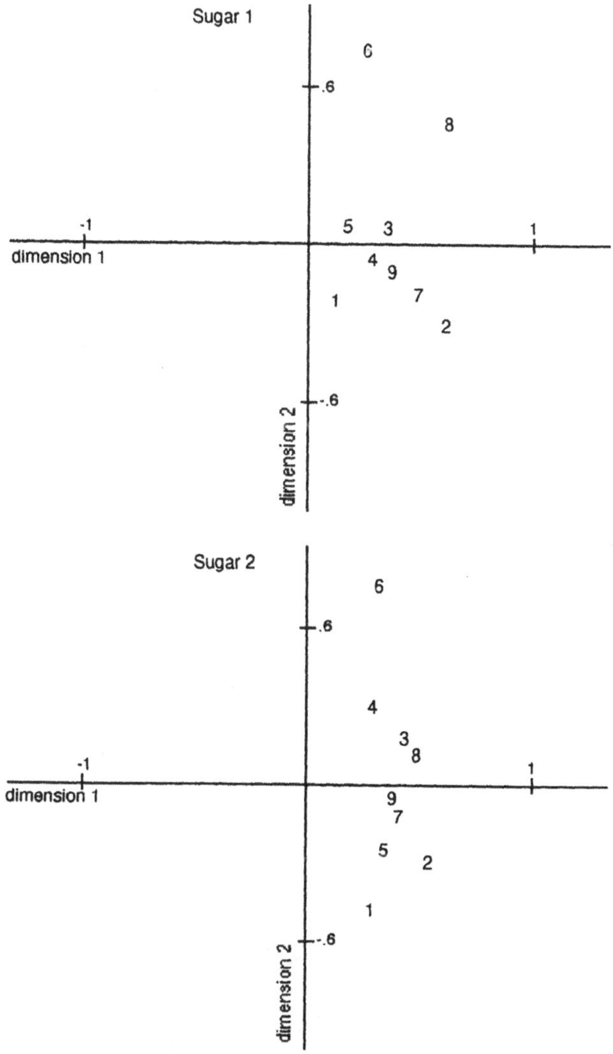

Figure 6 Loadings of the assessors on the first two Non-Centered Principal Curves (dimensions) for the two sugar concentrations.

Assessor 2 and 8 have a high loading on the first Non-Centered Principal Curves for both concentrations. The loading of assessor 1 is small, the TI-curve of assessor 1 does not comply very well with the shape and intensity of the first Non-Centered Principal Curves. Assessors 2 and 8 have TI-curves which are like the first Non-Centered Principal Curves.

Assessor 6 has the highest loading for the second Principal Curve, in contrast with assessors 1 and 2 which are among the lowest loadings on the second Non-Centered Principal Curve. Figure 3 shows the difference between the individual curves for the high-sugar concentration. Negative loadings on the second curve go together with slow rising and declining curves (e.g. assessors 1, 2 and 5), positive loadings mean fast rising and declining curves (e.g. assessor 6 and 8). The third Non-Centered Principal Curves seem to consist of mainly error and are not displayed for that reason.

11.4.2 Non-Centered PCA of TI Data From Four Bitter Solutions

Figure 7 shows the first Non-Centered Principal Curves of the four bitter solutions.

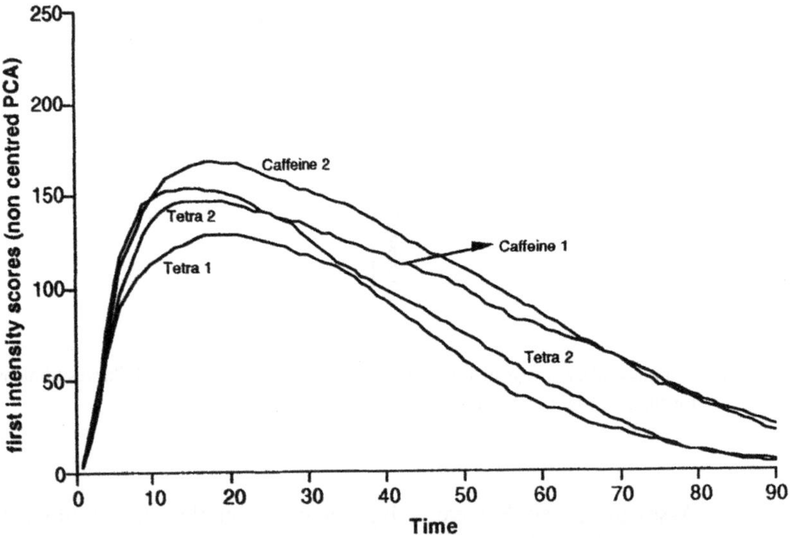

Figure 7 The first Non-Centered Principal Curves of the four bitter solutions.

The interpretation of the first Non-Centered Principal Curves of the bitter solutions is analogous to that of the sweet solutions. Again these curves look like the average curves. Figure 8 shows the second Non-Centered Principal Curves.

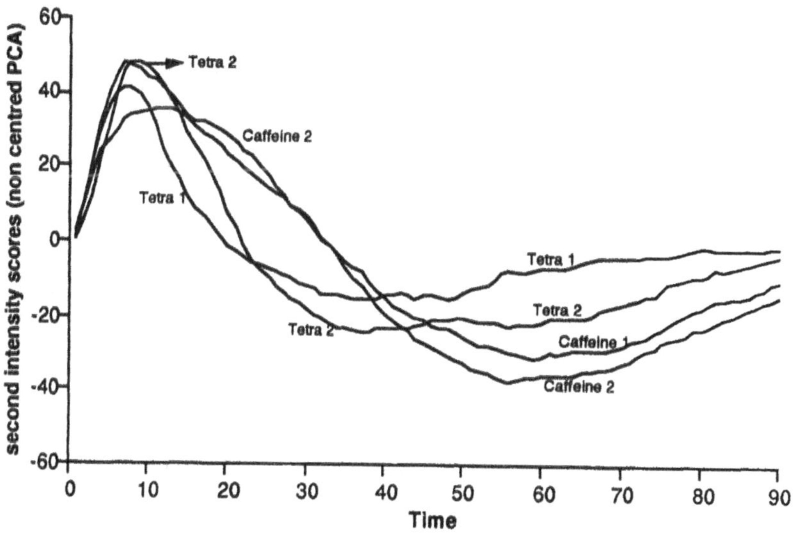

Figure 8 The second Non-Centered Principal Curves of the four bitter solutions.

The second Non-Centered Principal Curves for the bitter stimuli also turn out to model the rate of increase and decline of the original TI-curves. The loadings of the individual curves on the Principal curves again help in interpreting the Principal Curves. Figure 9 presents these loadings for the caffeine stimuli. Assessor 6 has a high loading on dimension 1 for the caffeine solutions, assessor 5 has low loadings. For the Tetrahop solutions assessors (loadings not shown) 2 and 4 load high and assessors 6 and 9 load low on the first Principal Curve. A negative loading on the second principal curve means a TI-curve with a slow rise and slow decline. A positive loading means a curve with fast rising and falling flanks. Another thing that can be seen from Figure 8 is that the two Tetra and the two caffeine stimuli cluster together from approximately 15 to 45 seconds. In the tail of these curves the clustering is less apparent.

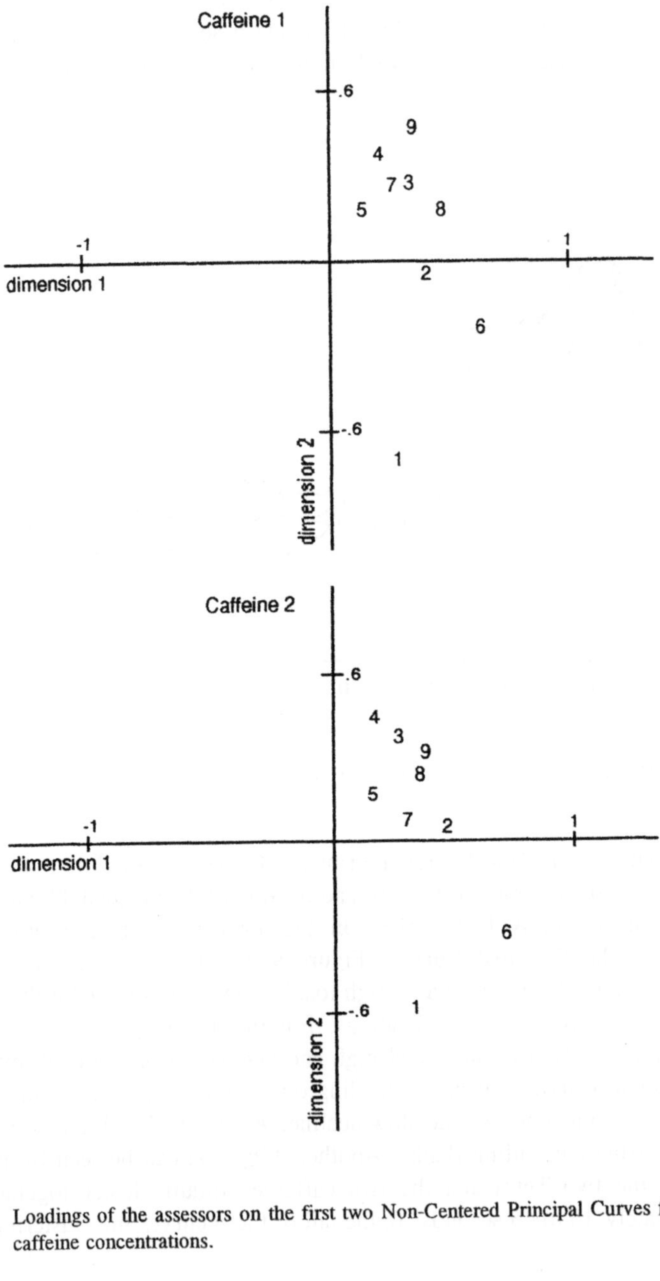

Figure 9 Loadings of the assessors on the first two Non-Centered Principal Curves for the two caffeine concentrations.

11.5 Covariance Principal Curves

Van Buuren (1992) suggested PCA on centered data. With this analysis the average is removed from the data and consequently PCA is applied to a matrix with covariances between the curves. In Figure 10 the first Covariance Principal Curves of the bitter solutions are shown.

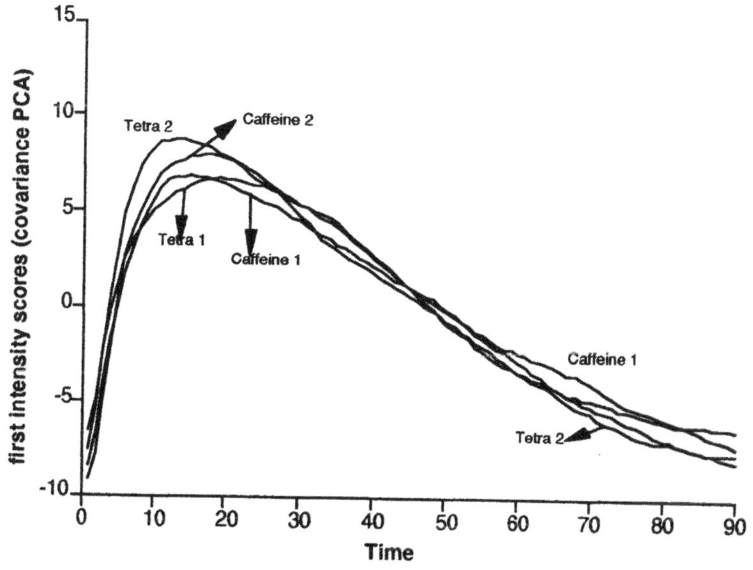

Figure 10 First Covariance Principal Curves for the bitter stimuli.

The first Covariance Principal Curves do not correspond with the average curves as the Non-Centered Principal Curves did. The expected differences in concentration are only visible at about the maximum of the first Covariance Principal Curves, not in the remainder of the curves. After 50 seconds the higher concentrations of the sweet solutions cross the lower concentrations. This is contrary to expectation since high concentrations usually have a longer aftertaste than the low concentrations.

Figure 11 shows the second Covariance Principal Curves for the four bitter stimuli.

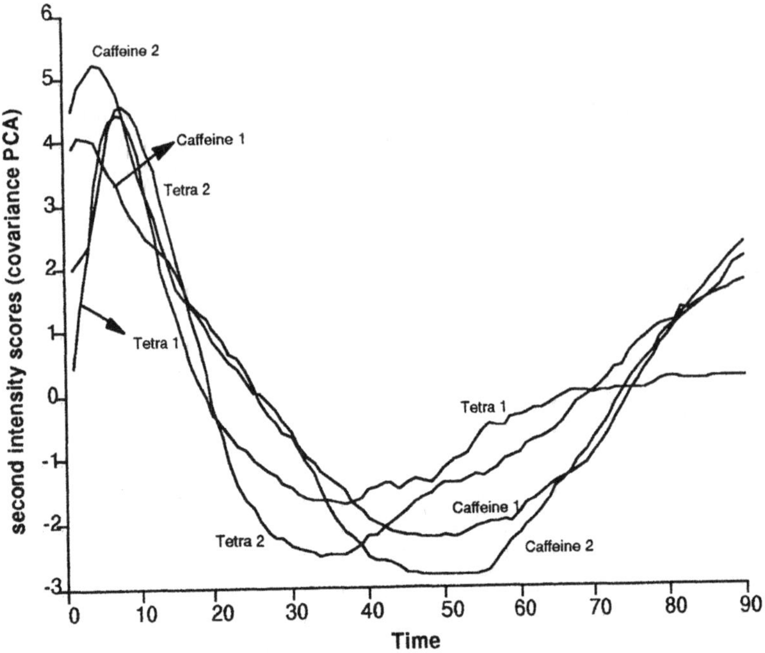

Figure 11 Second Covariance Principal Curves for the bitter stimuli.

Using the individual curves and curve loadings (both not shown) the Covariance Principal Curves can be interpreted. The first Covariance Principal Curves model the general shape of the TI-curve, though without overall height information, since the average curve is subtracted from the data. The second Covariance Principal Curves show the deviations in rise and decline from the first Covariance Principal Curves (as did the 2nd Non-Centered curves). The clustering together of the two Tetra and the two Caffeine curves can be seen in Figure 11 from approximately 20 to 70 seconds.

11.6 Correlation Principal Curves

When the data are corrected for the mean and divided by their standard deviation the principal curves resulting from the PCA are called Correlation Principal Curves. Now both overall height and individual spread is deleted from the data. What remains can be called the *general* shape of the curve. As a result the first Correlation Principal Curves (Figure 12) almost overlap each other because this *shape* of all TI-curves is generally the same. It is conceivable that the interesting information is in the variability and the level of the curves. Often level, and/or variability, are not interesting when studying shape, but with TI-curves this may not be the case. More research is needed to answer questions about the shape of TI-curves and which information can be extracted from TI-curves. For another definition of the of TI-curves see Dijksterhuis (1992a) and Dijksterhuis and van den Broek (1994).

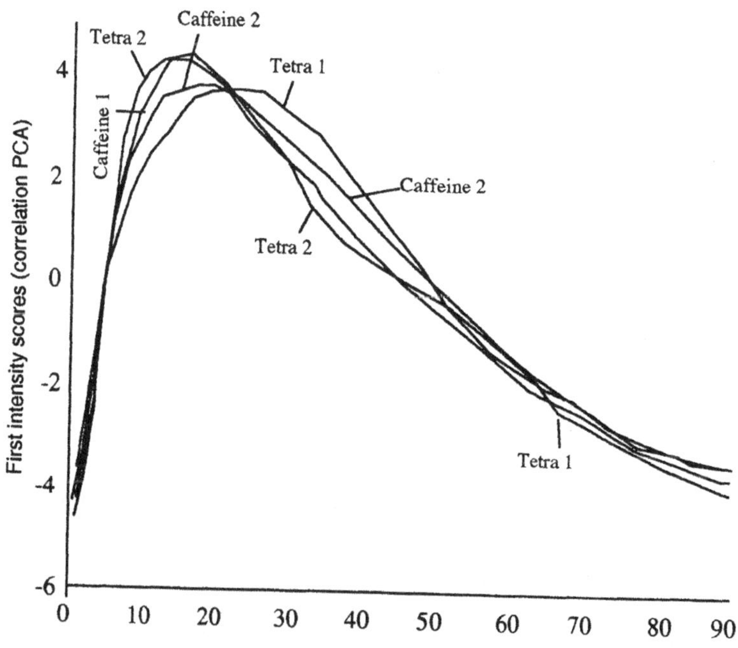

Figure 12 First Correlation Principal Curves for the bitter stimuli.

Differences in concentrations have disappeared in Figure 12. With help of the curve loadings (not shown) the first Correlation Principal Curves can be interpreted as the shape of the TI-curve in which time to maximum appears to play an important role. The second Correlation Principal Curves seem to model the deviation in time to maximum from the first Correlation Principal Curves (Figure 13).

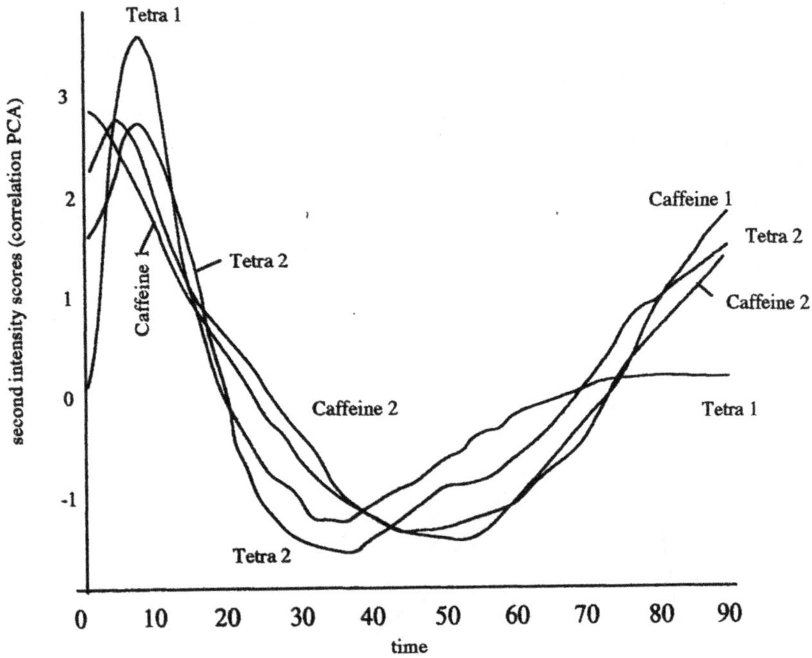

Figure 13 Second Correlation Principal Curves for the bitter stimuli.

The clustering together of the curves from the same stimulus can be seen in Figure 13 too. From approximately 20 to 70 seconds the two Tetra and the two caffeine curves cluster together.

11.7 Conclusion

Principal Component Analysis of TI-curves is a useful way of aggregating the TI-curves of individual assessors. The first Non-Centered Principal Component is the best aggregated TI-curve (of course under the assumptions of linear PCA). Second Principal Components can be interpreted and seem to model the rate of rise and fall of the TI-curves. The loadings of the PCA can be used to classify the assessors and to help in interpreting the obtained Principal Curves. The second Principal Curves show a clustering of the two Tetra and the two Caffeine stimuli, not of the sweet stimuli. This clustering is visible with all three PCA variants, perhaps somewhat more outspoken in the Covariance and the Correlation Principal Curves. This may point at some systematic difference between the time courses of the Tetra-taste and the Caffeine taste; further study could be interesting.

Correlation Principal Curves can be interpreted to reflect similarities in general "shape" of the TI-curves, whether or not enough information is in this "shape" remains to be investigated further. For aggregating all information in the TI-curves they are probably less useful, because level and variability information are deleted from the individual curves. Non-Centered Principal Curves model most aspects of the individual TI-curves. Covariance Principal Curves are in between Non-Centered and Correlation Principal Curves.

CHAPTER 12

Matching the Shape of Time-Intensity Curves

12.1 Introduction

Time Intensity (TI) research is used to study changes over time in the taste of food and beverages. The individual differences in the resulting Time-Intensity curves are often very large (see e.g. Overbosch *et al.* 1986, Pangborn *et al.* 1983). Often TI-curves are averaged over subjects, after which TI-parameters are obtained from the average TI-curves. Figure 1 shows an example of this simple averaging process for two curves.

By Garmt Dijksterhuis and Els van den Broek. Originally published in 1995 in Journal of Sensory Studies (1995), 10, 149-161. (Reprinted with kind permission from Food & Nutrition Press)

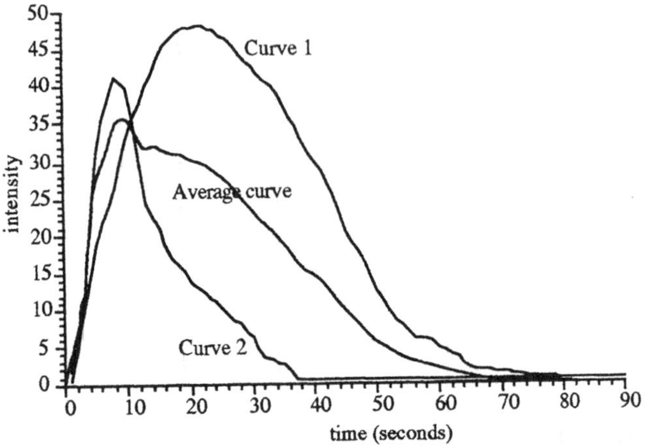

Figure 1 Example of averaging two TI-curves.

The question is whether the averaged TI-curve is representative of the individual curves. The averaged TI-curve often has properties unrepresentative for both curves (see e.g. Liu and MacFie, 1990). For the study in this paper we will ask: "Is the shape of the average curve representative of the *shapes* of the individual curves?"

Recently Principal Component Analysis was suggested for the analysis of TI-curves (van Buuren 1991, 1992, Dijksterhuis 1993, Dijksterhuis *et al.* 1994). In Principal Component Analysis (PCA) of TI-curves each individual curve receives a weight and a weighted average curve is computed. The weight works in the Intensity-direction only, the Time-direction remains untouched. A method that would also weight the time-direction would be interesting. The isotropic scaling in the shape-matching method works both in the time- and intensity-direction. The method is suggested here to provide another type of aggregating the TI-curves, and perhaps to point at a potential new direction of TI-curve data analysis.

12.2 Method: Shape Analysis

Usually TI-curves are regarded as vectors, which are one-dimensional. The vector contains the intensity values, one for each time-sample. When the TI-measurement took one and a half minute and the sample frequency was one per second, this vector is an array of 90 intensity values. A common way of inspecting the curves is by plotting the intensity values against time-values, with time on the abscissa and intensities on the ordinate (as in Figure 1). The plotted curve lies in a plane spanned by the dimensions *time* and *intensity*. In the remainder of this article we define the term shape as the configuration in two-dimensional *time-intensity* space (see also Dijksterhuis 1992). In this space the curves are characterised by pairs of numbers: (*time, intensity*). Each curve is now represented by an ($S \times 2$) matrix with S the number of time-samples. A well known method for matching two or more matrices is Generalised Procrustes Analysis (Gower 1975, Dijksterhuis and Gower 1991/2), but this method is not employed here because not all Procrustes transformations are useful for TI-data interpreted as curve-shapes.[1]

Procrustes *Rotations* are not used because the shapes as defined above have a unique orientation, the time- and intensity axes should remain the underlying dimensions of the space. *Translation* of the whole curve does not alter the shape of the curve, it merely shifts the curve, but it has an effect on the *isotropic scaling* which stretches or shrinks the curve. This translation shifts the curves around a common centre. A closer match of the curves is achieved by the subsequent isotropic scaling when the curves are centred (see Dijksterhuis 1992).

Isotropic scaling is the stretching or shrinking of the entire TI-curve in an equal amount in all directions of the space. Figure 2 gives an illustration of an isotropically stretched and a shrunk TI-curve. Stretching is achieved by multiplication of the entire curve with a scaling factor $\rho > 1$, shrinking by $0 < \rho < 1$. The method of isotropic scaling is described in Gower (1975), ten Berge (1977) and Peay (1988).

[1] When the data are arranged in a 3-mode structure (e.g. time-samples × stimuli × assessors) GPA may be useful for analysing TI-curves. This is however a different approach from the one used in this paper.

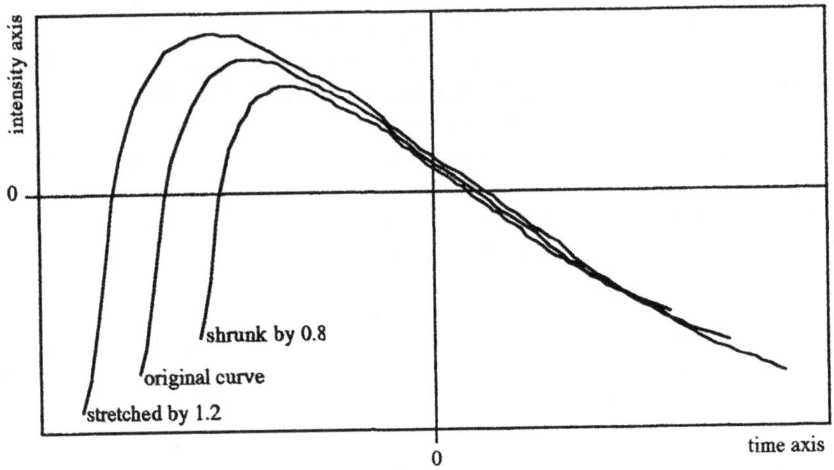

Figure 2 Illustration of an isotropically stretched and shrunk TI-curve.

Isotropic scaling as applied to a data-set of TI-curve shapes is performed on curves from different assessors and the same stimulus. This results in a separate analysis for each stimulus.

12.3 Examples

12.3.1 Bitter Solutions I

The results of a small study in which 9 subjects judged 8 bitter stimuli is analysed using the method proposed above. The stimuli were 2 concentrations (Low, 125 mg/l (L) and High, 160 mg/l (H)) of Caffeine (CL, CH, CLr, CHr; r: replicated) and 2 concentrations (12 mg/l and 15 mg/l) of TetraHop (tetra-hydro-iso-humulone, TL, TH, TLr, THr; r: replicated). The data are taken from Flipsen (1992, see also Dijksterhuis *et al.* 1994). For each stimulus there are 9 individual TI-curves. Replicates were treated as separate stimuli. For each stimulus a separate analysis is performed. The 8 separate analyses result in a (8×9) table of scaling weights (see Table 1).

Table 1 Scaling weights from the analyses of the 9 individual curves from the 8 bitter solutions from Flipsen (1992)

	1	2	3	4	5	6	7	8	9
CL	.9163	1.02	1.062	1.069	1.134	.8903	1.083	.8906	.9934
CLr	.7604	.8841	1.018	1.076	1.22	.9619	1.054	1.103	1.027
CH	.8453	1.05	.9773	.9884	1.146	.9774	1.073	1.108	.9379
CHr	1.02	.8929	.9575	1.062	1.087	.9816	1.07	.9786	.9885
TL	1.117	.9535	.9951	.9551	1.079	1.124	.9812	.8592	1.026
TLr	1.092	.8115	.8902	.9982	1.008	1.174	1.122	.9601	1.139
TH	1.041	.8849	1	.961	1.191	1.192	.9346	.9027	1.042
THr	.6589	.8454	1.065	1.038	1.162	1.228	.9978	.9879	.9748

It is hard to see the structure in the scaling weights. When the scaling weights were equal to 1 this would have meant that all curves were identical, no stretching or shrinking was necessary. The weights actually are close to 1, which is a result of the fact that all TI-curves have approximately the same shape: they rise to a maximum and slowly decline to zero. To find out whether the small differences in the scaling weights contain any underlying pattern the table was analysed by means of Principal Component Analysis (see e.g. Jolliffe 1986). A two-dimensional representation of the 8 stimuli resulted (see Figure 3). The first two dimensions explained 40.4% and 24.8% variance respectively. In Figure 3 can be seen that the Caffeine stimuli are better replicated than the TetraHop stimuli. The scaling weights for the two pairs of replicated Caffeine stimuli were close for most assessors. The shape of the curves of different Caffeine concentrations are closer than for TetraHop.

The loadings of the PCA on the data in Table 1 represent the 9 subjects, and are plotted in Figure 4. It looks like the subject-group was rather homogeneous, there are no clearly identifiable outliers or clusters of subjects.

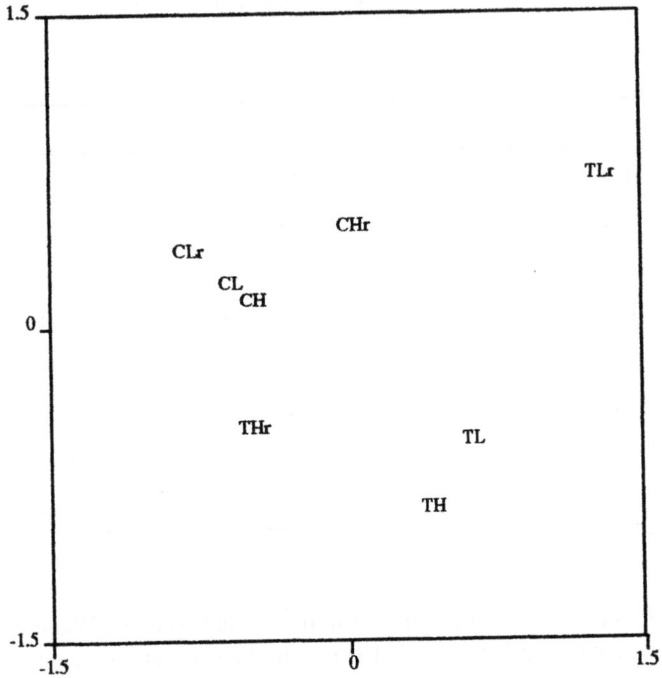

Figure 3 Two dimensional representation of the 8 bitter stimuli.

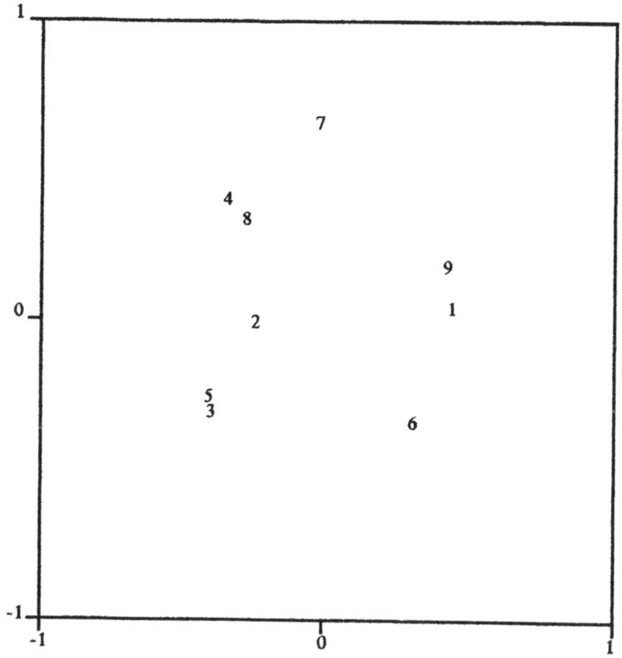

Figure 4 Position of subjects for the analysis of the bitter-solution TI-curves.

12.3.2 Bitter Solutions II

The second example concerns data from van den Broek (1993). Three concentrations (Low, 1250 mg/l (L); Medium, 1950 mg/l (M); High, 2650 mg/l (H)) of Caffeine (CL, CM, CH) and 3 concentrations of Quinine (125 mg/l (QL), 142.5 mg/l (QM), 160 mg/l (QH)) were analysed. Each stimulus was presented three times and the mean curve over replications was calculated. For each stimulus there are 14 individual mean TI curves. Each stimulus is analysed separately, this resulted in a (6×14) table of scaling weights (see Table 2).

Table 2 Scaling weights of the 6 analyses on the 14 individual curves from the 6 bitter solutions (2 tastes, 3 concentrations) from van den Broek (1993).

	1	2	3	4	5	6	7	8	9	10	11	12	13	14
CL	1.008	0.957	0.985	1.016	1.015	1.016	1.013	1.002	0.999	1.014	0.987	1.016	0.967	1.01
CM	1	0.986	0.989	1.015	1.011	1.015	0.992	1.005	1.007	1.014	0.983	1.013	0.969	1.004
CH	0.995	0.966	0.988	1.019	1.024	1.024	1.025	0.991	0.973	1.01	0.977	1.021	0.988	1.006
QL	0.945	0.942	0.963	1.05	1.006	1.057	1.053	0.921	1.049	1.016	1.027	1.052	1.038	0.93
QM	0.952	0.942	0.989	1.05	1.029	1.055	1.048	0.91	1.049	0.939	1.048	1.053	1.052	0.938
QH	0.924	0.942	1.022	1.051	1.035	1.07	0.969	0.975	1.051	0.978	1.028	1.066	1.056	0.891

To visualise the structure in the table PCA was applied to it. The results are plotted in Figure 5. The first dimension explained 69.1% variance, the second 19.5%.

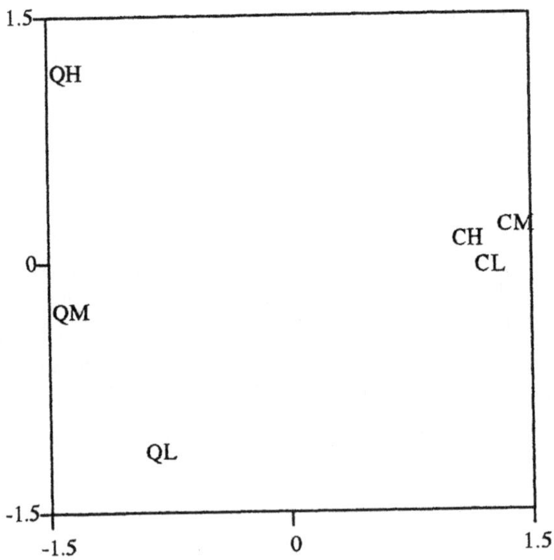

Figure 5 Structure of TI-curves of the Quinine and Caffeine stimuli.

In Figure 5 the curves of the different Caffeine stimuli are closer than the curves of the Quinine stimuli. This means that the differences within the set of the Caffeine stimuli are smaller than those in the set of Quinine stimuli. The first dimension seems to represent a taste component. The second dimension seems to be a concentration effect for Quinine, however for Caffeine there appears no clear concentration effect.

The group of subjects seems to divide into two small clusters and some loose subjects. On the left in Figure 6 there is a group of subjects 12, 6, 13, 4, 9 and 11, on the right we find 1, 14, 10 and 2. The pattern of scaling weights over the stimuli seems to differ for these two groups. Since the first dimension represents a taste-effect (Quinine versus Caffeine) it can be inferred that the leftmost cluster of subjects has received higher scaling weights for the Quinine curves. The rightmost cluster has higher weights for the Caffeine curves. Remembering that high weights correct for relatively small curves the leftmost group must have had small curves for Quinine. Analogously the right group has small curves for Caffeine.

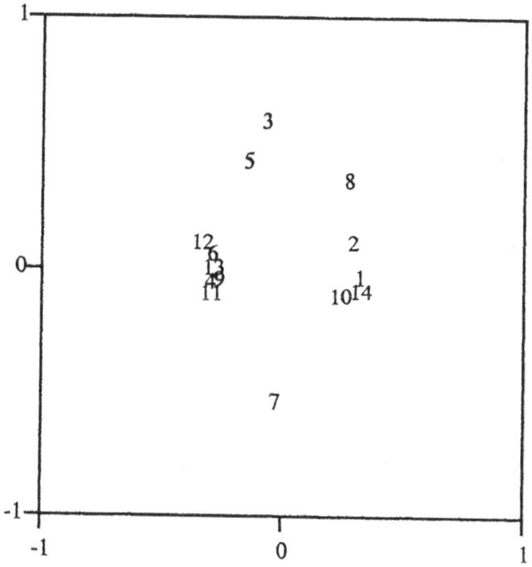

Figure 6 Position of subjects for the analysis of the bitter-solutions TI-curves.

It seems that two main segments can be identified in the group of subjects. The two segments systematically have a different shape of their TI-curves for Quinine and Caffeine stimuli.

Segmentation becomes interesting with larger groups of subjects, but a trend can already be found samples of smaller size.

12.3.3 Fat Spreads

In this example the data from 35 different fat-spreads are analysed (van den Broek 1993). The fat-spreads were made of 4 carriers, viz. one margarine (M), two kinds of low fat margarine (L1, L2) and an experimental spread (X). Each spread was prepared with three tastes, diacetyl (D), 3-heptanone (H), ethylheptanoate (E) in three concentrations (Low (L), Medium (M), High (H)). Each stimulus was evaluated three times by 14 judges. The mean curve over these 3 replicates was calculated for each subject. The analysis was done for each stimulus separately, so a (14×35) table of scaling weights is the result (see Table 3). To find structure in these weights a PCA was carried out. The first two dimensions explained respectively 31.8% and 16.7% variance (see Figure 7).

Table 3 Scaling weights of the 35 analyses (fat spreads, in rows) of the 14 individual curves (in columns).

	1	2	3	4	5	6	7	8	9	10	11	12	13	14
1	1.016	0.964	0.979	0.917	1.037	1.042	1.032	0.996	1.015	1.044	0.916	1.034	1.01	1.023
2	0.987	0.96	0.951	0.946	1.043	1.045	1.039	0.971	1.014	1.034	0.928	1.048	1.02	1.046
3	0.998	0.972	1.005	0.93	1.026	1.028	0.982	1.027	0.992	1.008	0.989	1.031	1.024	0.997
4	1.012	1.019	0.99	0.945	0.984	1.013	0.995	1.027	0.994	0.981	1.005	1.024	1.017	1
5	1.025	0.968	1.001	0.969	1.032	1.034	1.029	0.995	1	1.028	0.98	1.03	1.031	0.899
6	1.017	0.961	0.985	0.93	1.002	1.034	0.973	1.007	1.007	1.021	1.011	1.04	1.025	0.998
7	1.012	0.968	0.969	0.879	1.043	1.044	1.043	1.023	0.994	1.038	0.968	1.038	0.978	1.038
8	1.022	0.991	1.021	0.908	1.029	1.028	1.004	0.999	0.98	0.99	1.004	1.025	1.022	0.99
9	1.025	1.018	0.992	0.953	1.006	0.992	0.995	0.989	0.998	1.002	1.021	1.033	1.026	0.958
10	1.011	1.01	0.976	0.933	1.012	1.032	1.002	1.019	0.984	1.012	1.01	1.03	1.014	0.963
11	1.035	0.984	0.964	0.927	1.013	1.01	1.025	1.016	1.012	0.997	1.02	1.037	0.991	0.979
12	1.055	0.944	1.051	1.055	0.92	1.054	1.042	1.042	1.036	0.97	0.946	1.044	1.041	0.862
13	0.974	0.984	1.002	1.014	1.011	1.011	0.998	1.006	0.999	1.009	0.959	1.015	1.012	1.01
14	0.979	0.989	1	1.003	1.014	1.015	1.002	1.008	0.984	0.999	0.989	1.002	1.009	1.007
15	0.945	1.019	1	0.998	1.008	1.017	0.975	1.017	0.994	0.993	0.987	1.022	1.013	1.019
16	0.988	1.025	0.997	0.981	1.005	1.03	1.004	1.018	0.985	0.993	0.979	1.02	1.009	0.971
17	0.947	1.012	1.016	1.015	1.004	1.01	1.018	0.999	0.972	1.008	0.974	1.011	1.004	1.015
18	0.979	0.938	1.022	1.027	1.004	1.032	1.006	1.027	0.997	1.01	0.97	1.031	1	0.971
19	1.065	0.926	1.063	1.061	0.973	1.068	1.065	1.041	1.023	0.908	0.985	1.054	1.045	0.814
20	1	0.933	1.004	1.023	1.024	1.02	1.008	1.01	0.988	1.006	0.972	1.023	1.018	0.98
21	1.055	0.945	1.049	1.05	0.962	1.052	1.049	1.005	1.025	0.934	1	1.042	0.991	0.888
22	1	0.886	1.006	1.018	1.025	1.024	1.019	1	0.981	1.009	0.998	1.029	1.002	1.022
23	0.983	1.019	1.015	0.997	1.005	1.014	1	0.982	0.98	1.015	0.99	1.02	1.012	0.971
24	1.039	1.019	1.046	1.043	1.038	1.043	1.036	1.032	0.958	1.005	0.98	1.045	0.991	0.793
25	1.001	0.997	1.005	1.021	1.001	1.024	0.949	0.989	1.001	1.005	0.984	1.023	1.014	0.991
26	1.002	0.96	1.037	1.012	1.033	1.039	1.023	0.995	1.012	1.013	0.974	1.045	1.016	0.872
27	0.977	1.003	1.02	1.008	1.014	1.015	0.921	1	1.001	1.006	0.999	1.02	1.01	1.015
28	0.987	0.979	1.021	1.021	1.018	1.024	1.016	1.006	1.022	1.023	0.991	1.01	0.997	0.901
29	0.986	0.965	1.002	1.02	1.003	1.022	0.965	0.998	1.008	1.001	1.007	1.033	0.995	1
30	0.998	0.97	1.013	1.013	1.008	1.015	1.003	1.002	0.991	1.008	0.989	1.013	0.971	1.011
31	0.958	1.015	1.023	1.022	1.015	1.021	1.023	0.997	1.02	1.022	0.989	1	1.016	0.897
32	1.03	1.003	1.029	1.03	0.948	1.029	1.027	1.01	0.915	1.005	0.963	1.028	0.974	1.028
33	0.992	0.994	1.021	1.022	1.015	1.019	0.997	0.998	1.002	0.99	0.986	1.019	0.985	0.966
34	1.02	0.961	1.007	1.02	0.994	1.037	0.938	0.966	1.003	1.007	1.015	1.02	1.033	0.992
35	1.021	1.012	1.023	1.023	0.999	1.018	1.02	1.015	0.961	0.947	0.955	1.023	0.975	1.02

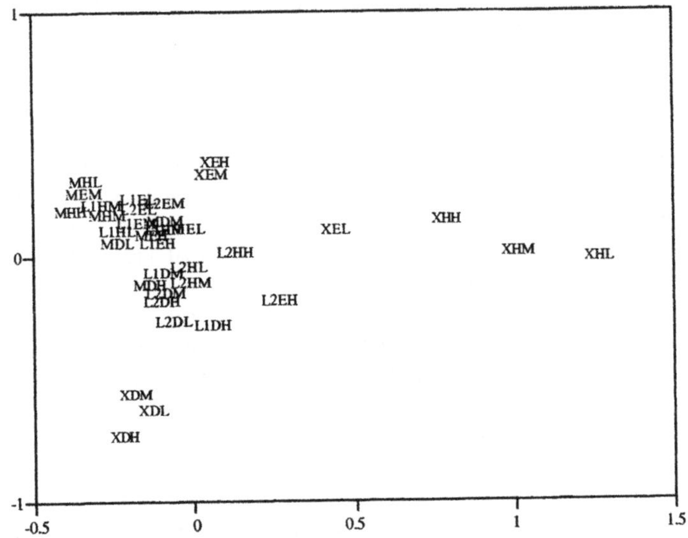

Figure 7 Structure of TI-curve shapes of the fat-spread stimuli.

In Figure 7 all experimental stimuli (X) are positioned differently from the other spreads (L1, L2 and M), irrespective of taste (XD, XH, XE) and concentration (-L, -M, -H). The other stimuli seem to cluster in two loose groups, one containing mainly margarine (M) and one of the low fat margarine (L1) stimuli, the other group mainly L2 stimuli. Furthermore there seems to be a taste component visible. The E tastes are mainly found in the upper leftmost group while the H tastes are distributed over the upper part of the plot. The D tastes are concentrated in the lower part of the figure. The effect of concentration is not clearly visible in the plot. The first dimension seems to represent a carrier component. A taste component (D versus E and H) can be found in the second dimension.

Some loose clusters of subjects can be seen in Figure 8. At the lower left subject numbers 14, 10 and 5, at the lower right part number 1, 6, 7, 9, 12 and 13 seem to form such a loose cluster. These clusters are clearly not as tight as the ones found in the previous example (Figure 5). Therefore it can be concluded that the 14 subjects in this experiment seems to have rather homogeneously evaluated the stimuli. No clear systematic differences in the shape of their curves seem to have occurred.

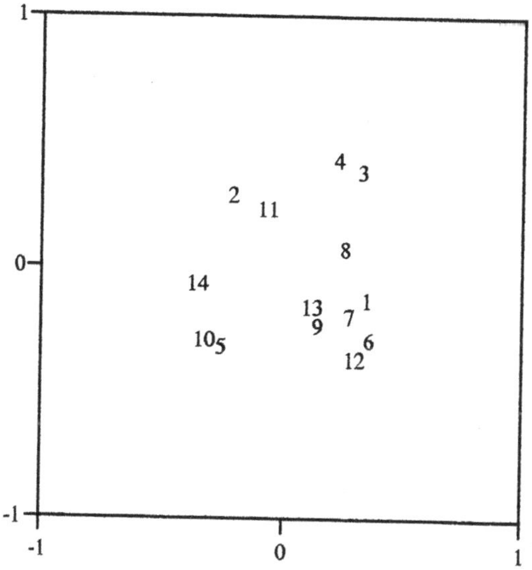

Figure 8 Position of subjects for the analysis of the fat-spread TI-curves.

12.4 Conclusion

Focusing on the *shape* of the TI-curve in two-dimensional "Time-Intensity- space" may be a starting point for some new methods of aggregating and analysing TI-curves (compare "shape theory", Stoyan and Stoyan 1990 or Goodall 1991). The results permit conclusions about shape-differences in the TI-curves resulting from different stimuli. The method also gives a plot which can be studied for segmentations of subjects and identification of outliers. The example analyses of the three TI-data sets in this paper lead to potentially interesting conclusions about the underlying TI-curves and segmentations of the group of experimental subjects. The configuration of stimuli may be a bit hard to interpret without some knowledge of the stimuli. The plot with the judges clearly shows outliers and groups of subjects scoring similarly.

Another method which immediately suggests itself is based on *non*-isotropic scaling. This means that the TI-curves are modified differently in the time- and in the intensity direction. Weight factors for the time- and intensity direction separately are thus obtained which may be studied as proposed.

Concluding Remarks Part IV: Time-Intensity Data Analysis

Summary

In this chapter the main conclusions of the chapters in Part IV are summarised and suggestions for future research are given.

Chapter 10
Principal Component Analysis of Time-Intensity Bitterness Curves

The Principal Component Analysis of Time-Intensity curves theoretically results in better aggregated Time-Intensity curves than the average curve over assessors. The interpretability of the Principal Curves is not straightforward and knowledge of the stimuli is needed. The non-centered Principal Component Analysis may give results that are better interpretable than the Principal Curves from other Principal Component methods, but a more systematic comparison of Principal Component methods is given in the next chapter. The Non-Centered Principal Time-Intensity Curves may suggest phenomena at certain points in time which can be investigated further in new experiments. As an exploratory tool the (non-centered) Principal Component (*Curve*) Analysis is promising.

Chapter 11
Principal Component Analysis of Time-Intensity Curves: Three Methods Compared

This chapter provides the necessary comparison of three Principal Component Analysis variants. The previous chapter was lacking such a comparison. There is some evidence for substantiation of the tentative conclusion from Chapter 10 ("Non-centered analysis is the most promising method."), but more research is needed.

Another aspect of the Principal Component Analysis of Time-Intensity curves is the interpretation of the loadings (weights). These are used to identify segments of assessors and outliers. They can also be useful to help interpreting the Principal Curves.

Suggestions for Future Research

In the previous two chapters replicated curves were averaged before the Principal Component Analysis was carried out. Replicated Time-Intensity curves, i.e. curves from the same assessor judging the same stimulus, usually are almost identical. When replicated curves are included in the matrix as extra columns, this gives the possibility to check the validity of the results, provided that the number of replications is the same for each stimulus. Ideally the Principal Curves of replicates should coincide, in practice they will be close. When they are far apart something has gone wrong in the study, an assessor is unreliable or another source of error is active.

To be able to judge the Principal Component method on its merits for Time-Intensity data analysis a large number of Time-Intensity data sets should be analysed. Preferably data sets that were already analysed by conventional means and of which the interpretation is clear and well established, should be used for these analyses.

The Principal Components could be attributed a significance measure by performing some kind of permutation- or randomisation test. This would extend the Principal Component Analysis-Time-Intensity method with a way to judge whether or not to interpret 2nd, 3rd or higher Principal Curves.

Another use of the Principal Component Analysis of Time-Intensity-curves is suggested by van Buuren (1991). Using only the first few Principal Curves, so-called Predicted Curves can be computed. These curves are smoothed versions of the original data because the higher components, in casu higher, Principal Curves are not included. These predicted curves can easily be constructed as linear, weighted, combinations of the original curves.

Reflection of Curves

In the Chapters 10 and 11 the computed Principal Curves are presented together in one plot. Each of the Principal Curves in the plot resulted from a separate analysis, so there is no relation between these curves. It can so happen that some curves appear upside-down. This is no problem usually with the first principal curves because they can just be plotted with the correct orientation. The reflection of a principal component is allowed. But with higher principal curves, i.e. second or third, their bumpiness increases and it can become hard to give them all an orientation for easy display. Since the comparison is between the curves in one plot, this is a matter of concern. It suggests not to take into account very bumpy curves, i.e. too high principal curves.

It should not be hard to devise a way to match the principal curves, to give them orientations as similar as possible. This could e.g. be done by comparing the squared distances between their corresponding time points, for both orientations, and picking the orientation which results in the lowest squared distance. This enterprise is left to other researchers.

Chapter 12
Matching the Shape of Time-Intensity Curves

This chapter contains a suggestion of a new direction of research of Time-Intensity curves. The underlying idea is still that Time-Intensity curves from the same stimulus can be matched in some way, while correcting for individual differences. The method suggested is one of many possible methods, and its advantage is that both Intensity- and Time-scores are weighted (Principal Component Analysis only weights Intensity-scores). The interpretation of the isotropic scaling weights computed is not straightforward, unfortunately. A Principal Component Analysis of the table of weights gives results that can be interpreted, but there is much room for improvement.

The plots with the assessors can be used to find segments or outliers, this is a useful property of the method, but alternative methods may be useful too.

Suggestions for Future Research

The method can be divided in two main steps:

1. estimation of the scaling factors
2. representation of the scaling factors by Principal Component Analysis

In Chapter 12, the latter is done both by a table and by a plot of the results of a Principal Component Analysis on this table. It is clear that with a small number of factors a table may suffice. But the number of estimated factors is KN, i.e. the product of the number of assessors and the number of stimuli. In most Time-Intensity experiments this number will be quite large.
The two steps above can be stated in more general terms:

1. transformation of the Time-Intensity curves
2. representation of the transformations in a convenient way

The first step should correct for individual differences between the Time-Intensity-curves of different assessors in order to match the shape of the Time-Intensity-curves, depending on the stimulus. This could mean that for each stimulus a separate analysis is performed as was done in the method suggested in Chapter 12.

For the first step a number of methods can be used of which anisotropic scaling was already mentioned. Another possibility is computing an $(N \times N)$ matrix of pairwise distance- or association measures between the Time-Intensity curves. This matrix can be analysed by an appropriate Multidimensional Scaling Method.

For the second step other Principal Component Analysis-like methods can be used that may be more convenient for the representation of the scaling factors resulting from step 1.

Future Directions in Time-Intensity Research[1]

Apart from Principal Component Analysis other Multivariate methods may be used to analyse Time-Intensity-data. Different approaches to Time-Intensity-data analysis are also conceivable. In this section some suggestions are given.

Multivariate Analysis of Time-Intensity Data

In addition to the above mentioned suggestions for future research some other Multivariate Analysis techniques may be interesting for the analysis of Time-Intensity data. For the Principal Component Analyses presented, two-mode matrices (time-points $\times$ assessors) are analysed for each stimulus separately.

[1] Paul Eilers (DCMR, Milieudienst Rijnmond, The Netherlands) is thanked for his comments concerning the "Curve fitting" approach. Dr. Paul Lewi (Janssen Pharmaceutical Company, Beerse, Belgium) is thanked for suggesting the compartment model.

Three mode matrices (time-points × stimuli × assessors) may be analysed by 3-mode methods. An obvious method to start with is Generalised Procrustes Analysis (see footnote 1 in Chapter 12), but other three-mode methods may be used too.

Another way to add a third mode in the analysis is by including replicates. Replicates may even be added as a fourth mode to the above mentioned three-mode data, giving (replicates × time-points × stimuli × assessors) data. But the analysis of four-mode data is very complex and there are very few methods available for this. One might consider MANOVA methods, but the assumptions of MANOVA will almost certainly be violated by the Time-Intensity data. More research is needed here.

Curve Fitting

Another approach to analysing Time-Intensity-curves is by fitting them onto theoretical curves. A family of functions that seems able to describe the shape of Time-Intensity curves reasonably well is suggested by Eilers (1993a):

$$\hat{y}_{ij}=a_j(1-e^{-b_j x_{ij}})e^{-c_j x_{ij}} \tag{1}$$

where i indexes the time-points and j the assessors. The parameters a_j, b_j and c_j are estimated iteratively using the method of "projected curves". This method is so far only applied experimentally to a small Time-Intensity data set (Eilers 1993b), but it may be an interesting direction for further research.

Pharmacokinetic Approach

In pharmacology the path a pharmacon takes through the body, from ingestion to the receptors at the site where it is needed, can be modelled by a system of equations known as the compartment-model. The pattern of activity of the pharmacon, plotted against time, looks like a Time-Intensity-curve. The compartment-model may be an interesting model to study Time-Intensity-curves. After all the path of a pharmacon is comparable to that of taste-stimulus: saliva, mucus, receptor-site at the tongue.

CHAPTER 13

Concluding Remarks

13.1 Introduction

It is not easy to formulate general conclusions concisely given the diversity of statistical methods employed. One problem is the apparent schizophrenic character of this book. It is on sensory research, but by fits and starts it appears to be on data analysis, without which, by the way, sensory research cannot be conduced properly. The Concluding Remarks, at the end of each part, reflect the schism by giving separate conclusions for sensory science and data analysis. This schizophrenia may be rationalised by keeping in mind that this is really a book on sensory research and not on data analysis. The data analysis is a means to analyse the data. That some of the Multivariate techniques are introduced rather mathematically results from the cooperation with their developers, who are mathematicians. All techniques are introduced elsewhere in considerable more detail and appropriate references are provided. The focus here is on their applications to sensory data, or what is called *Sensometrics*. Given that Sensometrics is an evolving field of research, there is no schism really.

Often a particular problem of -sensory- data analysis calls for modifications or extensions of MVA methods. But these are secondary to the main topic of this book: the analysis of data from sensory and consumer science: Sensometrics.

In the following, each part will be presented in turn, their main conclusions summarised and indications for future research given.

13.2 PART I: Individual Differences

Individual differences between assessors in a sensory or consumer panel are acknowledged to exist. Despite this, averaging over assessors, which is almost never justified, is widely practised. In Chapter 2 it is shown that even in a conventional profiling panel the use of the attributes can be so different that this casts doubt on any analysis that averages attributes over assessors. Only one-dimensional attributes can be averaged safely, and almost no attribute appears strictly one-dimensional.

There are large differences between panels, and some panels will have attributes that are more one-dimensional -have higher consonance values- than others. It would be interesting to find out what are the causes of the differences between panels. Is it a matter of the extent or method of training of the assessors? Does it result from the particular set of products? Does it result from panel fatigue; was the number of presentations too high? These are matters worth investigating.

There may be an optimum, somewhere between maximal consonance and minimum experimentation costs. This optimum may perhaps not be obtained by sensory research at this moment, and it can certainly not be achieved by averaging data over assessors. Finding this optimum could lay at the base of a research program in sensory- and consumer-panel research.

Another conclusion from the low consonances encountered *must* be that individual difference models, such as GPA, must be applied to panel data. When GPA is applied it is in principle not important whether one uses classical GPA (Gower 1975) or Projecting GPA (GPPA, Peay 1988). The idea is of course that an individual difference model is applied, and GPA is the simplest and most well known method.

Chapter 3 presents both GPA and GPPA and illustrates GPPA by analysing both a conventional profiling data set and a free choice profiling data set. Once the GPPA or GPA method is applied, using different optimisation criteria hence giving difference results, the different fit and loss variance measures from a GPPA can be interpreted just as the corresponding measures from a GPA. Chapter 3 presents these measures in a large number of tables and bar-charts. (For a somewhat more detailed comparison of GPA and GPPA, and some other individual difference models, see Dijksterhuis and Gower (1991/2)). An important point that is made in this chapter, as in the previous chapter, is that GPA (or GPPA) is a mandatory method for FCP data, but perhaps should be considered mandatory for conventional profiling data too.

13.2.1 Individual Differences Analysed by GPA

Different methods for GPA (coined "classical" GPA and GPPA in Chapter 3) fit different models and give different interpretations of the data. Though in practice these differences may be small, it is important to realise that differences exist (see also Dijksterhuis and Gower 1991/2). These differences can be linked to the underlying research question of the sensory experiment at hand.

The question

> *Can you match the individual data matrices as well as possible, i.e. using all available similarity between the individual data matrices?*

is answered positively by the Classical Procrustes Analysis (GPA) method (Gower 1975). This method uses all dimensions, overlooking no information. The result is a solution of maximum dimensionality, say R, of which the researcher has to choose a low number r for the representation and interpretation of the results.

The question

> *Can you plot as much information from the data in a low number of dimensions, e.g. two dimensions to make graphical representations?*

can be answered affirmatively when one uses the Projecting Procrustes Analysis (GPPA) method (Peay 1988). With this method there is an a priori choice of dimensionality and the fitting of the model takes place irrespective of any similarity that may exist in higher dimensions. As a result the solutions are not nested, i.e. a p-dimensional solution is not the same as the first p dimensions of a $(p+1)$-dimensional solution. When the researcher has an a priori reason to look for an r dimensional structure, and wants to find this structure best, though at the risk of overlooking information in higher dimensions, GPPA is the method to use.

It is impossible to give a conclusion as to which method is best used, it depends on the question asked, and on preferences of the researcher.

13.2.2 Some Methodological Issues

The results of GPA are commonly presented as two-dimensional plots of the configurations of products, sometimes superimposed with plots of the loadings or correlations of the attributes. Further application of linear biplot theory (Gabriel 1971) in conjunction with GPA may lead to visually appealing and better interpretable configurations. The GPA Group Average could be biplotted with the structure correlations of the attributes from the different assessors (compare Ter Braak, 1990, see also Dijksterhuis and Gower 1991/2). This may enable easy interpretation of the use of the attributes, the properties of the objects and, if wished, reification of the dimensions.

The so-called "assessor-plots" sometimes found in connection with GPA (see e.g. Arnold and Williams 1985) are a device that deserve further investigation. Though they are not presented in this book, a few points should be addressed. The plots give a graphical representation of the assessors in a panel. This representation is the result from a classical scaling, or Principal Coordinate Analysis (PCO, Gower 1966, 1984) of a symmetric matrix containing the Procrustes loss values between the individual configurations of each pair of assessors (Banfield and Harries 1975, see also Harries and MacFie 1976).

The comparison of assessor-plots based on the losses of GPA solutions of different dimensionality may be a way to help find an optimal dimensionality to represent the results. It may also serve as a means to identify outliers or segments in the panel, and perhaps find assessors or segments responsible for a particular dimension in the solution.

A lot of sensory data may be more easily collected as ordinal or nominal data, instead of as numerical data. The abundance of numerical data does not mean that there really is information in the numbers. Ordinal or nominal (or mixed) data may contain just as much information. A non-linear GPA method, without isotropic scaling and only applied experimentally once (van Buuren and Dijksterhuis 1988), exists and it would be worthwhile to revitalise the study of this method.

13.2.3 A Misconception about GPA

In the latest printing of their textbook on sensory research, Stone and Sidel (1993) share a frequently held misconception about GPA. This is very unfortunate because their book is considered a standard textbook on sensory evaluation.

Stone and Sidel appear to address two different points:

- QDA is not so time-consuming as is stated by some
- GPA is not a good method to analyse profiling data

Their first point is correct. Often Free Choice Profiling was presented as an alternative to QDA and other conventional profiling methods. Though it is true that FCP can be applied with minimal training -hence fast- it is a matter of discussion whether this is any good. Their second point is incorrect. They state about FCP data:

> The individual results and the panel results are then analysed using a generalised Procrustes analysis. This type of factor analysis analyses data until an acceptable result is achieved; it is a highly experimental procedure, one that should not be considered without full appreciation for the consequences. (Stone and Sidel, p. 237)

The procedure is said to analyse "data until an acceptable result is achieved", but this is exactly what most statistical and data-analytical methods do, and no reason to doubt the method. Their concern seems to be that GPA "is a highly experimental procedure", which may have been true in 1962 when the method was introduced, or in 1975 when the Generalisation of two-sets Procrustes Analysis was published. After almost 20 years of applying GPA to sensory data and further research into the method it cannot be maintained that GPA is "a highly experimental procedure". That one should not consider any statistical method without appreciation for the consequences is true. But this can be said for the averaging over assessors too!

Stone and Sidel continue:

> As Huitson (1989) noted, the analysis always produces a result that might not be justified based on examination of the database itself. (Stone and Sidel, p. 237)

The article by Huitson (1989) does not show anything with respect to the justification of Procrustes results (see the reaction by Arnold (1990) and the answer of Huitson (1990) to this). Huitson used configurations of random data to show that GPA can make sense out of nothing. Unfortunately he only performed one random data analysis for a particular problem. Had he performed a larger number of random data analyses, say 50, he would have performed a random-data test for his particular GPA data set, and could have found a significant difference between "real" data and random data. The conclusion would have probably been that GPA is able to differentiate between "real" data, i.e. containing "signal", and random data containing only "noise".

13.3 PART II: Measurement Levels

Chapter 4 presents a flexible method for the analysis of K data sets which can be applied to the data from the assessors in a sensory or consumer panel, as was illustrated by a very small image study. In Chapters 5 and 6, non-linear GCA is presented for the analysis of sensory data sets.

13.3.1 Distances as Models for Sensory Perception

The analyses in Chapter 4 are rather complicated, which may inhibit their rapid adoption by others. However, their general formulation suggests a paradigm for a direction of fundamental sensory research. The method was summarised as follows:

$$\mathbf{X}_k \xrightarrow{\text{distance}} \mathbf{D}_k \xrightarrow{\text{map}} \mathbf{Y}_k \xrightarrow{\text{match}} \mathcal{M}(\mathbf{Y}_k) = \mathbf{Y} \qquad (1)$$

Its distance measures are reminiscent of the early developments in Multidimensional Scaling and allied methods. In fact the approach to first compute distance matrices and later match the mapped distances was the rationale for the PINDIS system of Lingoes and Borg (1978, see also Commandeur 1991). MDS methods were applied to construct so-called *perceptual spaces*, which could be interpreted as a model of the way individuals, or a group of individuals, perceived a set of stimuli. It may be interesting to study distance models to represent a set of sensory stimuli in an optimal way. Of course this has been done to a certain extent (by e.g. Schiffman *et al.* 1981), using a number of MDS programs, but the current proposal is not restricted to any existing MDS method or computer program.

A number of questions must be answered first:

1. Which distance measure is best for presentation of the stimuli?
2. Is there an optimal mapping of the distances into perceptual spaces?
3. What is the optimal way to match the individual spaces?

The first question concerns the choice of the distance-generating function, coined g in Chapter 4. Some methods may use aggregated data:

$$g(\mathbf{X}) = \mathbf{D} \qquad (2)$$

where $\mathbf{X}$ is e.g. the average of the individual $\mathbf{X}_k$. Of course an individual difference approach may be more promising:

$$g(\mathbf{X}_k)=\mathbf{D}_k \tag{3}$$

Some assessors may need other distance functions than others:

$$g_k(\mathbf{X}_k)=\mathbf{D}_k \tag{4}$$

specifying a different distance-generating function g_k for each assessor. Possibly some variables may need other distances than other variables. This situation was encountered in Chapter 4, but alike for all assessors. Adding another subscript generalises this so that each variable for each assessor may be analysed using a different distance:

$$g_{jk}(\mathbf{X}_{jk})=\mathbf{D}_k \tag{5}$$

The different distances generated are, for each assessor, combined into one distance matrix $\mathbf{D}_k$. The problem is to choose the appropriate distance functions g_{jk}. Experience, theory, common sense, experimentation, etc. may give suggestions.

The next step in the scheme (1) is the mapping. Once different functions g_{jk} are found, the same mapping method can be used for all assessors. Whether this will be PCO or another method depends on the particular data set. Note that in Chapter 4 the combination of the Extended Matching Coefficient for g and PCO defines Multiple Correspondence Analysis, and choosing Euclidean distances with PCO defines Classical Scaling.

The individual representations $\mathbf{Y}_k$ can be studied, preferably after giving them similar orientations by means of GPA. The preferred mapping method in the proposed paradigm is GPA because of the rigid-body-character of its transformations: distances are preserved fixed in the matching.

13.3.2 Individual Differences Again

The application of Generalised Canonical Analysis (GCA), as an alternative to Generalised Procrustes Analysis (GPA), enables a detailed study of sensory data. Because it is an individual difference model individual assessors can be identified and studied, just as with GPA.

In Chapter 5, GCA is shown to be useful for the analysis of Multiway data, i.e., 3-way data from sensory research. Even when the original line-scale scores are recoded into a mere three categories, there emerges an interpretable solution. A distinction between butcher made and factory made sausages is found, just as van Buuren (1987) had found using a different method. In Chapter 6, GCA is applied to another sensory data set, and two different solutions are compared. These two solutions, from an ordinal 3-category analysis and from a numerical 10-category analysis, are almost identical.

The important point is that, again, an individual difference model is employed, and that in addition non-linearities can be modelled. The latter is necessary when analysing categorical data on an nominal or an ordinal measurement level. In addition non-linear relationships between variables are modelled too (see Chapter 1, §1.6.2 and §1.6.3).

13.3.3 Future Sensory Research

Additional research into the use of response scales is needed, for example a systematic study of the differences in responses of assessors to:

1. a low number of categories on the scale used
2. a line-scale to indicate an intensity

Usually line-scales are used, but they may only provide mock information in their numerical values. Category-scales may be easier to use by the assessors, resulting in more consistent use of the scale and less fatigue. There may be an optimum number of categories, perhaps depending on the type of product. On the other hand the assessor may feel restricted by the categories and may prefer to give a mark along a continuous line-scale.

Studying the differences between category and line-scales means performing the same experiment twice, once under condition 1 and once under condition 2. Such an experiment is different from a *post hoc* recoding of categories as in Chapter 5 and 6, though the results could turn out equivalent. However, the experiment should be performed first.

Using non-linear methods of MVA is appropriate for the analysis of the data from the experiments suggested above. When an ordinal analysis of a number of categories shows linear transformations of the category-numbers, one could confine oneself to linear methods for subsequent analyses.

Identifying nonlinearities may be useful in understanding the responses of the assessors, or to select a number of variables with certain properties. Comparing a numerical, an ordinal and a nominal analysis can reveal which variables behave in a non-linear manner. These variables can be further inspected or deleted from subsequent ordinal or numerical analyses if such analyses are thought more appropriate.

It may be easier for assessors to assign one of a relatively low number of categories, (for an attribute) to a product, than to assign a number to the intensity of the attribute (see also van Buuren 1987). The marking of a line-scale, as is common practice, implies a judgement of the distance of the mark to the left and right end of the scale, or to the anchors. The perceived intensity of an attribute must be converted into a position on a line scale, hence into a distance to the end of the scale, or to an anchor. The task for the assessor looks like a kind of cross-modality matching, which is not an easy task in general. The assignment of a limited number of fixed categories may be easier for the assessor and result in data with less error. The analysis of this kind of qualitative data is best performed with non-linear analyses, e.g. using ordinal PCA or GCA.

13.3.4 Data Analysis

There is a fundamental difference between GPA and GCA in the way the matching of the individual data sets is performed. With GPA distances between products are kept fixed for each individual data set $\mathbf{X}_k$, with GCA the distances change as a result from the projections involved. It would be interesting to know how exactly GPA relates to GCA and which position is taken by GPPA.

The extension of non-linear GPA and its relation with non-linear GCA would be an interesting line of research (see also §13.2.2).

The same misconception about GPA, noted in §13.2.3 will probably arise for some people, concerning GCA instead of GPA. Though there is less experience with applying GCA than in applying GPA there is no need for a reserved -conservative- attitude towards the use of GCA for sensory data. Re-read §1.9 to find out why.

13.4 PART III: Sensory-Instrumental Relations

Sensory-Instrumental data usually consists of two data matrices. One data set contains scores on sensory attributes for a number of products, and the other set contains chemical/physical measurements on the same products.

There is no reason why Sensory-Instrumental relations should be linear, so it is an obvious choice to use non-linear statistical methods. In part III two of these methods are used: nonlinear Redundancy Analysis and nonlinear Canonical Analysis. The linear method employed in Chapter 9, PCA, proves useful in screening both the instrumental and sensory data sets. GPA (linear) can be used as a first approximation for the match of an instrumental and a sensory data set. When the fit of this match is low, nonlinear relations can be suspected. Then it may be time to use nonlinear methods, such as used in Chapter 7 and 8.

When the sensory set contains hedonic judgements ("appreciation", "liking", "quality", etc.) there is a large probability of nonlinear responses. So nonlinear methods are preferred in such cases (see e.g. Noble 1975). Performing both linear and nonlinear (e.g. ordinal) analyses using RA or CCA enables one to identify the most nonlinear attributes.

A disadvantage of nonlinear methods, especially of (nonlinear) CCA is the possibility of obtaining a very high, but artificial, fit. This can result from the freedom of the optimal scaling algorithm to choose nonlinear transformations of the variables together with a capitalisation on correlations between pairs of variables, one in each set. An almost perfect fit is always suspect, and one should look for the reason. Nonlinear CCA is sensitive for extreme categories with low frequencies. When they are there, they can be recoded or the corresponding variable deleted. It will be useful to perform another analysis with a somewhat more strict measurement level. Finally a linear method may be applied, such as GPA, as suggested in Chapter 9. The balance between the use of non-linear and linear methods may be hard to obtain. Knowledge of the products under investigation, and of the statistical methods, should result in a reasonable choice of analysis method and hopefully a reasonable balance.

Though two data sets are involved, this is not a necessary restriction. GCA (*K*-sets CCA) can be applied instead of CCA, and GPA can already handle more than two sets. In Chapter 7 and 8 three data sets are related: Design variables, Instrumental variables and Sensory variables. Instead of matching the sets pairwise they could very well be analysed using GPA or GCA.

13.5 PART IV: Time-Intensity Data Analysis

TI-data consist, as it were, of mini time-series of just one or two minutes length. Two different ways of exploring the structure present in the TI-curves are presented in Part IV. In Chapter 10 and 11 PCA is used to this end, and different variants of PCA are compared. The non-centered analysis may give the best results because both level and spread information are retained, which are important properties of the TI-curves, and may reveal interesting psychological or physiological differences among assessors. Different products may differ in just these features. Chapter 12 suggests another approach, that of the analysis of the shape of the TI-curve. This is a highly experimental approach, which can only be seen as the first step in a new exploratory direction to analyse TI-curves. As mentioned in Chapter 12, in addition to the isotropic scaling of the curves, non-isotropic scaling may be applied, or other ways of fitting the curves. The Concluding Remarks of Part IV suggest some other Multivariate Analyses of TI-curves.

The MVA methods applied in the chapters 10, 11 and 12 are mainly exploratory methods. In addition to the development of the exploratory analyses of TI-curves it may be timely to start thinking of other approaches. The curve-fitting of the TI-curves onto theoretical curves, and the use of the compartment model from Pharmaco-kinetics, are two methods suggested in the Concluding Remarks to Part IV.

13.5.1 Theory Driven Approach

Time-Intensity research is quite different from the other research presented in this book. In a way it is closer to the nose and, in particular, tongue, than the other three parts in this book. TI-curves are a representation of an assessor's perceived intensity of a stimulus over a short period of time. This kind of research is close to the original psychophysical research of Weber, Fechner and Stevens. But the development on the understanding of the time-course does not reflect this closeness. Little is known about the exact underlying processes that give the TI-curve its shape. Not many people doubt that there is something to be found in the curves, but all agree that it is obscured by individual differences, and perhaps other influences. The three chapters in Part IV try to model the individual differences by combining individual TI-curves in some way or another. The resulting pooled TI-curves may be interpreted further, but this is where much TI-research stops. Stimuli have different TI curves, some have longer after-taste, some have long lasting high overall intensity, some build up the taste very slowly, some build up taste very quickly. It is time that the conclusions of TI-research go beyond the descriptive.

There is a lot of knowledge about the perception of taste-stimuli. Much of this knowledge is of a chemical, biological and physiological nature. In the "Understanding Flavour Quality" (1992) symposium some of the latest developments were presented (see the corresponding issue of Food Quality and Preference, 5, no.1 and 2). There is also psychophysical knowledge about the processes of smelling and tasting. There are theories about higher processes of perception, though usually for vision and not for taste and smell. This knowledge may be studied for suggestions for developing new theories of smell and taste. All this may lead to new ways of studying and interpreting TI-curves.

13.6 Closing Remarks

The diversity of the research presented in this book has been grouped into four main topics in Sensometrics:

- Individual Differences
- Measurement Levels
- Sensory-Instrumental Relations
- Time-Intensity Data Analysis

Each ideally will profit from the analyses presented. But, as will be clear from all the "suggestions for future research", there are no definite solutions yet. The chapters in this book indicate directions of research in which a potential solution may be concealed, and must be interpreted as such.

A general conclusion may be that the research field coined *Sensometrics*, is an interesting and lively field, and that there may be much to be gained from the research taking place in it. It is hoped that this book will contribute in the forwarding of sensometric research and that some of the suggested future research will actually be carried out, initially in an academic environment.

REFERENCES

Arnold, G.M. (1986). A Generalised Procrustes Macro for Sensory Analysis. *Genstat Newsletter*, 18, 61-80.

Arnold, G.M. (1990). Problems with Procrustes analysis. *Journal of Applied Statistics*, Letters to the Editor, 17, 449-451.

Arnold G.M., Williams A.A. (1985). The use of generalised Procrustes analysis in sensory analysis. In *Statistical Procedures in Food Research* J.R. Piggott (ed.). Amsterdam: North Holland.

Banfield, C.F., Harries, J.M. (1975). A technique for comparing judges' performance in sensory tests. *J. Fd. Technol.*, 10, 1-10.

Benzécri, J.P. (1973). *l'Analyse des Données (Data Analysis)* (2 vols). Paris: Dunod.

Buja, A., Eyubogly, N. (1992). Remarks on Parallel Analysis. *Multivariate Behavioural Research*, 27, 509-540.

CAMO Inc. (1992). *Unscrambler II, v4.0/3.7. User's guide.* Trondheim: CAMO A/S.

Carroll, J.D. (1968). Generalisation of canonical correlation analysis to three or more sets of variables. *Proceedings of the 76th Annual Convention of the American Psychological Association*, 227-228.

Carroll, J.D. (1972). Individual differences and multidimensional scaling. In R. N. Shepard, A.K. Romney, S.B. Nerlove. (Eds.), *Multidimensional scaling. Theory and Applications in the Behavioural Sciences.* Volume I: Theory. New York: Seminar Press.

Carroll, J.D., Arabie, P. (1980). Multidimensional scaling. *Annual Review of Psychology*, 31, 607-649.

Cliff, N. (1966). Orthogonal rotation to congruence. *Psychometrika*, 31, 33-42.

Cliff, N. (1987). *Analysing Multivariate data.* Florida: Harcourt Brace Jovanovich.

Cochran, W.G., Cox, G.M. (1957). *Experimental Designs*. Chichester: Wiley.

Commandeur, J.J.F. (1991). *Matching Configurations*. Leiden: DSWO Press.

Coppi, R., Bolasco, S. (1989). *Multiway Data Analysis*. Amsterdam: North-Holland.

Cramwinckel, A.B. and Van Mazijk-Bokslag, D.M. (1990). Een Chemische, microscopische en sensorische evaluatie van veertig merken groentesoep *(A chemical, microscopic and sensory evaluation of forty brands of vegetable soup.* In Dutch). RIKILT report 90-28. Wageningen: RIKILT-DLO.

Davies, P.T., Tso, M.K. (1982). Procedures for reduced-rank regression. *Applied Statistics*, 31, 244-255.

Davis, A.W. (1978). On the asymptotic distribution of Gower's m^2 goodness-of-fit criterion in a particular case. *Ann. Inst. Statist. Math.*, 30, 71-79.

de Leeuw, J. (1986). *On the history of redundancy analysis*. Research note. Leiden: Department of Data Theory.

de Leeuw, J. (1973). *Canonical Analysis of Categorical Data*. Thesis University of Leiden. (Reissued 1985. Leiden: DSWO Press).

Dember, W.N., Warm, J.S. (1979). *Psychology of Perception*. Holt-Saunders International Editions.

de Wijk, R.A. (1989). *Temporal factors in human olfactory perception*. Unpublished thesis, University of Utrecht, the Netherlands.

Dijksterhuis, G.B., (1987). *Gegeneraliseerde Niet-Lineaire Canonische Correlatie Analyse: toepassingen in het sensorisch onderzoek.* (Generalised non-linear Canonical Correlation Analysis: applications in Sensory Research. In Dutch) Unpublished master's, University of Utrecht, the Netherlands.

Dijksterhuis, G.B. (1990/1993). Een correctie van een sessie-effect door Procrustes rotatie van de correlatie-configuraties van de gemiddelde set (Unpublished internal report. Utrecht: OP&P) / Correcting a Session-Effect by means of Procrustes Rotations (Unpublished draft).

Dijksterhuis, G.B. (1991). (Unpublished internal report). Utrecht: Oliemans Punter & Partners.

Dijksterhuis, G.B. (1992a). Matching Time-Intensity Curves. In: *Proceedings of the "3émes jounees Europeennes, Agro-Industrie et Methodes Statistiques."* 30-11/1-12-1992, Monpellier, France. 139-142.

Dijksterhuis, G.B. (1992b). Response to Comments on Scriven and Mak. *Journal of Sensory Studies*, 7, 81-82.

Dijksterhuis, G.B. (1993). Principal Component Analysis of Time Intensity Bitterness Curves. *Journal of Sensory Studies,* 8, 317-328.

Dijksterhuis, G.B. (1994). Procrustes Analysis in studying sensory-instrumental relations. *Food Quality and Preference*, 5, 115-120.

Dijksterhuis, G.B. (1995). Assessing Panel Consonance. *Food Quality and Preference*, 6, 7-14.

Dijksterhuis, G.B., Broek, E. van den (1995). Matching the Shape of Time-Intensity curves. *Journal of Sensory Studies,* 10, 149-161.

Dijksterhuis, G.B., Buuren, S. van (1988). *Procrustes-PC Version 1.0 Manual*. Utrecht: OP&P.

Dijksterhuis, G.B., Buuren, S. van (1989) *Procrustes-PC v2.0 User-Manual*. Utrecht: OP&P.

Dijksterhuis, G.B., Buuren, S. van (1990). *Procrustes-PC v2.1 Manual*. Utrecht: OP&P.

Dijksterhuis, G.B., Flipsen, M., Punter, P.H., (1994). Principal Component Analysis of Time Intensity data. *Food Quality and Preference*, 5, 121-127.

Dijksterhuis, G.B., Gower, J.C. (1991/2) The interpretation of Generalised Procrustes Analysis and allied methods. *Food Quality and Preference*, 3, 67-87.

Dijksterhuis, G.B., Heiser, W.J., King, B., Punter, P.H. (1994). Meeting report: First Sensometrics meeting *Food Quality and Preference*, 5, 233.

Dijksterhuis, G.B., Heiser, W.J. (1995). Multivariate Analysis and Experimental Design in Sensory Research. *Food Quality and Preference*, 6, 263-270.

Dijksterhuis, G.B., Kraakman, H., Buuren, S. van (1991). *Procrustes-PC Version 2.2. User Manual*. Utrecht: OP&P.

Dijksterhuis, G.B., Krabbe, P. (1991). (Unpublished internal report.) Utrecht: OP&P.

Dijksterhuis, G.B., Punter, P.H. (1990). Interpreting generalised Procrustes analysis 'Analysis of Variance' tables. *Food Quality and Preference*, 2, 255-265.

Dijksterhuis, G.B., Punter, P.H., Williams, A.A. (1992). Comments on Scriven and Mak. *Journal of Sensory Studies*, 7, 77-83.

Dijksterhuis, G.B., Roos, P.E. (1990). *Collection and Processing of Time-Intensity Data*. Paper presented at the Nordic Workshop "Advanced Sensory Analysis III. The Consumer in Focus". 30-1/1-2-1990, Aarhus, Denmark.

Edgington, E.S. (1987). Randomization Tests. New York: Marcel Dekker, Inc.

Eilers, P.H.C. (1993a). Esitmating Shapes with Projected Curves. Manuscript. DCMR Milieudienst Rijnmond, The Netherlands.

Eilers, P.H.C. (1993b). Ervaringen met het modelleren van tijd-intensiteit-curven. (Experiences with modelling Time-Intensity curves. In Dutch). Unpublished Research note. DCMR Milieudienst Rijnmond, The Netherlands.

Ennis, D.M., (1991). *Probabilistic models of perception*. Dissertation Agricultural University of Wageningen.

Flipsen, M. (1992). *Tijd Intensiteit Onderzoek*. I: Algemeen; II: Zoet-en bitteroplossingen (Time Intensity Research. I: General; II: Sweet and bitter solutions, In Dutch). Research report OP&P Utrecht/Agricultural University of Wageningen.

Frijters, J.E.R. (1980). *Psychophysical and psychometrical models for the triangular method*. Dissertation, Univerisity of Utrecht.

Gabriel, K.R. (1971). The biplot graphical display of matrices with application to principal component analysis. *Biometrika*, 58, 453-467.

Geladi, P. (1988). Notes on the history and nature of partial least squares (PLS) modelling. *Journal of Chemometrics*, 2, 231-246.

Gifi, A. (1981). *Nonlinear Multivariate Analysis*. Leiden: University of Leiden, Department of Data Theory.

Gifi, A. (1990) Nonlinear Multivariate Analysis. Chichester: J. Wiley and Sons.

Goodall, C. (1991). Procrustes Methods in the Statistical Analysis of Shape, *J. R. Statist.* Soc. B, 53, 285-339.

Gower, J.C. (1966). Some distance properties of latent root and vector methods used in multivariate analysis. *Biometrika*, 53, 3&4, 325-338.

Gower, J.C. (1971). Statistical methods for comparing different multivariate analyses of the same data In J.R. Hodson, D.G. Kendall, P. Tautu (Eds.), *Mathematics in the Archaeological and Historical Sciences*. Edinburgh: Edinburgh University Press.

Gower, J.C. (1975). Generalized Procrustes analysis, *Psychometrika*, 40, 33-51

Gower, J.C. (1984). Multivariate analysis: ordination, multidimensional scaling and allied topics. In: *Handbook of Applicable Mathematics: Vol. VI, Statistics* (ed. E.H. Lloyd). Chichester: J. Wiley and Sons, 727-781.

Gower, J.C. (1992a). Biplot geometry. RR-92-02 Leiden: Department of Data Theory.

Gower, J.C. (1992b). Generalized Biplots. *Biometrika*, 79, 475-493.

Gower, J.C., Dijksterhuis, G.B. (1992). Coffee images: A study in the simultaneous display of multivariate quantitative and categorical variables for several assessors. RR-92-06, Department of Datatheory, University of Leiden.

Gower, J.C., Dijksterhuis, G.B. (1994). Coffee images: A study in the simultaneous display of multivariate quantitative and qualitative variables for several assessors. *Quality and Quantity*. 28, 165-184.

Gower, J.C. and Harding, S.A. (1988). Nonlinear Biplots. *Biometrika*, 75, 445-455.

Green, B.F. (1952). The orthogonal approximation of an oblique structure in factor analysis. *Psychometrika*, 17, 429-440.

Greenacre, M. (1984). *Theory and Applications of Correspondence Analysis*. New York: Academic Press.

Greenacre, M.J. (1988). The limitations of multiple correspondence analysis, *Computational Statistics Quarterly*, 3, 249-256.

Guttman, L. (1941). The quantification of a class of attributes: a theory and method of scale construction. In P. Horst et al. (Eds.), *The Prediction of Personal Adjustment*. New York: Social Science Research Council.

Hand, D.J. and Taylor, C.C. (1987). *Multivariate Analysis of Variance and Repeated Measures*. London: Chapman and Hall.

Harries, J.M., MacFie, H.J.H. (1976). The use of a rotational fitting technique in the interpretation of sensory scores for different characteristics. *J. Fd. Technol.* 11, 449-456.

Heiser, W.J. (1992). *Sensory observations: Univariate or Multivariate Data?* Presentation at the first Sensometrics Group meeting, 23-24 July 1992, Oegstgeest, the Netherlands.

Heiser, W.J., Meulman, J.J. (1993). Nonlinear methods for the analysis of homogeneity and heterogeneity. RR-93-03, Department of Datatheory, University of Leiden.

Hirst, D., Muir, D.D., Næs, T. (1993). Definition of the organoleptic properties of hard cheese: a collaborative study between Scottish and Norwegian Panels. *International Dairy Journal*. 4, 743-761.

Hoaglin, D.C., Mosteller, F., Tukey, J.W. (1991). *Fundamentals of exploratory Analysis of Variance*. Chichester: Wiley.

Horst, P. (1961). Relations among m sets of measures. *Psychometrika*, 26, 129-149.

Hotelling, H. (1936). Relations between two sets of variates. *Biometrika*, 28, 321-327.

Huitson, A. (1989). Problems with Procrustes analysis. *Journal of Applied Statistics*, 16, 39-45.

Huitson, A. (1990). Comments on: 'Problems with Procrustes analysis' by Gillian M. Arnold. *Journal of Applied Statistics*, Letters to the Editor, 17, 451-452.

Hurley, J.R., Cattell, R.B. (1962). The Procrustes Program: producing direct rotation to test a hypothesized factor structure. *Behav. Sci.* 7, 258-262.

Israëls, A.Z. (1984). Redundancy analysis for qualitative variables. *Psychometrika*, 49, 331-346.

Israëls, A.Z. (1986). Interpretation of redundancy analysis: rotated vs. unrotated solutions. *Applied Stochastic Models and Data Analysis*, 2, 121-130.

Izenman, A.J. (1975). Reduced-rank regression for the multivariate linear model. *Journal of Multivariate Analysis*, 5, 248-264.

Johnson, R.M. (1963). On a theorem stated by Eckart and Young. *Psychometrika*, 28, 3, 259-263.

Jolliffe, I.T. (1986). *Principal Component Analysis*. Springer-Verlag.

Kettenring, J.R. (1971). Canonical analysis of several sets of variables. *Biometrika*, 56, 433-451.

King, B.M. (1986). Odor intensity measured by an audio method. *Journal of Food Science*, 5, 1340-1344.

King, B.M., Arents, P. (1991). A statistical test of consensus obtained from generalized Procrustes Analysis of sensory data. *Journal of Sensory Studies*, 6, 37-48.

Kjølstad, L., Isaksson, T., Rosenfeld, H.J. (1990). Prediction of sensory quality by near infraread reflectance analysis of frozen and freeze-dried green peas. *J. Sci. Food Agric.*, 51, 247-260.

Köster, E.P. (1971). *Adaptation and cross-adaptation in olfaction*. An experimental study with olfactory stimuli at low levels of intensity. Unpublished thesis, University of Utrecht, the Netherlands.

Koppenaal, E. (1991). Project: Objectieve meting voor detectie en voorspelling van de sensorische kwaliteiten stevigheid en meligheid in appellen. (Project: Objective measurements for detecting and predicting the sensory qualities firmness and mealiness of apples. In Dutch) Internal Report ATO-DLO Institute for Agrotechnology.

Kristof, W., Wingersky, B. (1971). *Generalisation of the orthogonal Procrustes rotation procedure to more than two matrices*. Proceedings of the 79th Annual Convention of the American Psychological Association, 6, 89-90.

Kruskal, J.B., & Shephard, R.N. (1974). A nonmetric variety of linear factor analysis. *Psychometrika*, 39, 123-157.

Krzanowski, W.J. (1984). Sensitivity of Principal Components. *J.R. Statis. Soc. B*, 46, 558-563.

Langeheine, R. (1982). Statistical evaluation of measures of fit in the Lingoes-Borg Procrustean individual differences scaling. *Psychometrika*, 47, 427-442.

Langron, S.P., Collins, A.J. (1985). Perturbation theory for generalized Procrustes analysis, *J.R. Statist. Soc. B*, 47, 277-284.

Larson-Powers, N., Pangborn, R.M. (1978). Paired Comparison and Time-Intensity measurements of the sensory properties of beverages and gelatins containing sucrose or synthetic sweeteners. *Journal of Food Science*, 43, 41-46.

Law, H.G., Snyder, C.W.Jr., Hattie, J.A., McDonald, R.P. (1984). *Research Methods for Multimode Data Analysis*. New York: Praeger.

Leach, E.J., Noble A.C. (1986). Comparison of bitterness of caffeine and quinine by time-intensity procedure. *Chemical Senses*, 11, 339.

Lebart, L., Morineau, A., Warwick, K.M. (1984). *Multivariate Descriptive Analysis*. New York: Wiley.

Lee III, W.E., Pangborn, R.M. (1986). Time-Intensity: The Temporal Aspects of Sensory Perception. *Food Technology*, November 1986, 71-82.

Leemans, H.M.J.M., Dijksterhuis, G.B., Trijp, J.C.M. van (1992). Positioneringsonderzoek van literaire boeken met Procrustes Analyse (Positioning of literary books using Procrustes Analysis. In Dutch), *Jaarboek 1992/1993 van de Nederlandse Vereniging van Marktonderzoekers*.

Lingoes, J.C., Borg, I. (1978). A direct approach to individual differences scaling using increasingly complex transformations. *Psychometrika*, 43, 491-519.

Liu, Y.H. (1990). *Multivariate Methods for Relating Different Sets of Variables; with applications in food research*. Ph.D. thesis, University of Bristol.

Liu, Y.H., MacFie, H.J.H. (1990). Methods for averaging Time-Intensity curves. *Chemical Senses*, 15, 471-484.

MacFie, H.J.H. (1992). Response to Comments on Scriven and Mak. *Journal of Sensory Studies*, 7, 82.

MacFie, H.J.H., Bratchell, N., Greenhoff, K., Vallis, L.V. (1989). Designs to balance the effect of order of presentation and first-order carry-over effects in hall tests, *Journal of Sensory Studies*, 4, 129-148.

MacFie, H.J.H., Liu, Y.H. (1992). Developments in the Analysis of Time-Intensity Curves. *Food Technology*, 46, February, 92-97.

MacFie, H.J.H, Thomson, D.M. (1984). Multidimensional Scaling Methods. In J.R. Piggott (Ed.), *Sensory Analysis of Foods*. England: Elsevier.

McBride (1990). Three generations of sensory evaluation. In: McBride, R.L., MacFie, H.J.H. (Eds.). *Psychological basis of sensory evaluation*. Elsevier Applied Science.

Meilgaard, M., Civille, G.W., Carr, B.T. (1990). *Sensory evaluation practices*. Boca Raton, Florida: CRC Press.

Næs, T., Solheim, R. (1991). Detection and interpretation of variation within and between assessors in sensory profiling. *Journal of Sensory Studies*, 6, 159-177.

Næs, T., Kowalski, B. (1989). Predicting sensory profiles from external instrumental measurements. *Food Quality and Preference*, 4/5, 135-147.

Neilson, A.J. (1957). Time Intensity studies. *Drug and cosmetic Industry*. 80, 452-453.

Nishisato, S. (1980). *Analysis of Categorical Data: Dual Scaling and its Applications*. Toronto: University of Toronto Press.

Noble, A.C. (1975). Instrumental Analysis of the Sensory Properties of Food. *Food Technology*, 29, December 1975, 56-60.

OP&P (1988). *PROCRUSTES-PC Version 1.0*. Utrecht: OP&P.

OP&P (1989). *PROCRUSTES-PC Version 2.0*. Utrecht: OP&P.

OP&P (1990). *PROCRUSTES-PC Version 2.1*. Utrecht: OP&P.

OP&P (1991). *The PSA-system v1.61. Manual*. Utrecht: OP&P.

OP&P (1992). *PROCRUSTES-PC version 2.2. A personal Computer Program for Generalised Procrustes Analysis*. Utrecht: OP&P Software Development.

Oreskovich, D.C., Klein, B.P., Sutherland, J.W. (1991). Procrustes Analysis and Its Applications to Free-Choice and Other Sensory Profiling, In Lawless, H.T., Klein, B.P. (Eds.) *Sensory Science Theory and Applications in Foods*. 353-393, New York: Marcel Dekker.

Oud, J.H.L., Blokland-Vogelesang, R.A.W. van (1993). *Advances in longitudinal and multivariate analysis in the behavioural sciences*. Nijmegen: ITS.

Overbosch, P., Enden, J.C. van den, Keur, B.M. (1986). An improved method for measuring perceived intensity/time relationships in human taste and smell. *Chemical Senses*, 11, 331-338.

Pangborn, R.M., Lewis, M.J., Yamashita, J.F. (1983). Comparison of time-intensity with category scaling of bitterness of iso-α-acids in model systems and in beer. *J. Inst. Brew.*, 89, 349-355.

Peay, E.R. (1988). Multidimensional rotation and scaling of configurations to optimal agreement. *Psychometrika*, 53, 199-208.

Piggott, J.R. (1993). Conference report: Understanding flavour quality: relating sensory to chemical and physical data. *Trends in Food Science & Technology*, 3, 329-330.

Punter, P.H, Roos, P.E., Dijksterhuis, G.B. (1989). *Measurement of the temporal aspects of sensory perception: Time-Intensity methodology.* 126th E.S.O.M.A.R. Seminar on 'Research for flavours and fragrances', Lyon, France, June 1989.

Punter, P.H. (1991). Sensorisch onderzoek: een historisch overzicht (Sensory research: a historical review, In Dutch) In C. de Graaf, P.H. Punter (Eds.), *Sensorisch onderzoek, eindelijk volwassen?*

Punter, P.H. (1992). Time intensity measurements, what do they tell us about perception? Presentation at the "Understanding Flavour Quality" symposium. *Understanding Flavour Quality. Relating Sensory to Chemical and Physical Data*. 20-23 September 1992, Bristol, England.

Rousset, S., Wakeling, I., Martin, J.F., MacFie, H.J.H., Touralle, C. (1992). Sensory profile and acceptability scores of different restructured steaks. In *Proceedings of the 38th International Congress of Meat Science and Technology*. August 23-28, 1992, Clermont-Ferrand, France, 273-276.

Schiffman, S.S., Reynolds, M.L., Young, F.W. (1981). *Introduction to Multidimensional Scaling*. Academic Press, New York.

Schlich, P. (1989). A SAS/IML Program for Generalised Procruses Analysis, In: 'Seugi '89', *Proceedings of the SAS European Users Group International Conference*, Cologne, May 9-12, SAS Institute GmbH, 529-537.

Schlich, P. (1993a). GRAPES: A method and a SAS program for Graphical Representations of Assessor Performances. *Journal of Sensory Studies*, 9.

Schlich, P. (1993b). Uses of Change-over designs and repeated measurements in sensory and consumer studies. *Food Quality and Preference*, 4, 223-235.

Schlich, P. (1993c). *Contributions à la Sensometrie* (Contributions to Sensometrics, partly French, partly English). Thesis University of Paris-sud, centre d'Orsay.

Schmitt, D.J., Thompson, L.J., Malek, D.M., Munroe, J.H. (1984). An improved method for evaluating time-intensity data. *Journal of Food Science*, 49, 539-542.

Schönemann, P.H. (1966). A generalized solution of the orthogonal Procrustes problem. *Psychometrika*, 31, 1-10.

Schönemann, P.H. (1968). On two sided orthogonal Procrustes problems. *Psychometrika*, 33, 19-33.

Schönemann, P.H., Carroll, R.M. (1970). Fitting one matrix to another under choice of a central dilation and a rigid motion. *Psychometrika*, 35, 245-255.

Scriven, F.M., Mak, Y.L. (1991). Usage behavior of meat products by Australians and Hong Kong Chinese: A comparison of Free Choice and Consensus Profiling. *Journal of Sensory Studies*, 6, 25-36.

Scriven, F.M. (1992). Response to Comments on Scriven and Mak. *Journal of Sensory Studies*, 7, 81.

Shepard, R.N, Romney, A.K., Nerlove, S.B. (1972). *Multidimensional Scaling. Theory and applications in the behavioural sciences*. Volume I: Theory. Seminar Press

Sibson, R. (1978) Studies in the robustness of multidimensional scaling: Procrustes statistics. *J. Roy. Statist. Soc.B.* 40, 234-238.

SPSS Inc. (1990). *SPSS Categories*. SPSS Inc.

Steenkamp, J.-B.E.M., Trijp, J.C.M. van, Berge, J.M.F. ten (1994). Perceptual Mapping based on idiosyncratic sets of attributes. *J. of Marketing Research*, Vol. XXXI, 15-27.

Stevens, S.S. (1960). The psychophysics of sensory function. *American Scientist*, 48, 226-254.

Stevens, S.S. (1962). The surprising simplicity of sensory metrics. *American Psychologist*, 17, 29-39.

Steyer, R., Wender, K.F., Widaman, K.F. (1993). *Psychometric Methodology*. Proceedings of the 7th European Meeting of the Psychometric Society in Trier. July 29-31, 1991, Germany.

Stone, H., Sidel, J.L. (1985). *Sensory Evaluation Practices*. first edition. Academic Press, Inc.

Stone, H., Sidel, J.L. (1993). *Sensory Evaluation Practices*. second edition Academic Press, Inc.

Stoyan, D., Stoyan, H. (1990). A Further Application of D.G. Kendall's Procrustes Analysis. *Biometrica Journal*, 32, 293-301, Berlin: Akademie-Verlag.

Tatsuoka, M.M. (1988). *Multivariate analysis. Techniques for educational and psychological sciences* (second edition). New York: Macmillan Publishing Company.

ten Berge, J.M.F. (1985). On the relationship between Fortier's simultaneous linear prediction and Van den Wollenberg's redundancy analysis. *Psychometrika*, 50, 121-122.

ten Berge, J.M.F. (1977). Orthogonal Procrustes rotation for two or more matrices. *Psychometrika,* 42, 267-276.

ten Berge, J.M.F., Knol, D. (1984). Orthogonal rotations to maximal agreement for two or more matrices of different column orders. *Psychometrika*, 49, 49-55.

ter Braak, C.J.F. (1990). Interpreting canonical correlation analysis through biplots of structure correlations and weights. *Psychometrika*, 55, 519-531.

Thomson, D.M.H., Williams, A.A., Wren, J.J. (1988). To build bridges. *Food Quality and Preference*, 1, 1.

Thomson, D.M.H. (1988). *Food Acceptability*. Elsevier Applied Science.

Thurstone, L.L. (1947). *Multiple-Factor Analysis. A development and Expansion of The Vectors of Mind*. Chicago: The University of Chicago Press.

Tyssø, V., Esbensen, K., Martens, H. (1987). UNSCRAMBLER, an Interactive Program for Multivariate Calibration and Prediction. *Chemometrics and Intelligent Laboratory Systems*, 2, 239-243.

"Understanding Flavour Quality" symposium. (1992). *Understanding Flavour Quality. Relating Sensory to Chemical and Physical Data*. 20-23 September 1992, Bristol, England.

van Buuren, S. (1987). Using multiple correspondence analysis in sensory quality research. *Chemistry and Industry*, 447-450.

van Buuren, S. (1991). *Principal curves of Time-Intensity data*. PRM 02-91, Leiden Psychological reports, psychometrics and research methodology. Department of psychology, Leiden University.

van Buuren, S. (1992) Analyzing Time-Intensity Responses in Sensory Evaluation. *Food Technology*, 46, November, 101-104.

van Buuren, S., Dijksterhuis, G.B. (1988). Procrustes analysis of discrete data. In M.G.H. Jansen, W.H. van Schuur (Eds.), *The Many Faces of Multivariate Analysis*, Proceedings of the SMABS-88 Conference, Vol. 1. Groningen: RION, Institute for Educational Research, 53-66.

van de Geer, J.P. (1984). Linear relations between k sets of variables. *Psychometrika*, 49, 79-94.

van de Geer, J.P. (1986). Introduction to linear multivariate analysis. Volume 1. Leiden: DSWO Press.

van den Broek, E. (1993). *Vergelijking van analyse-methoden voor Tijd-Intensiteit data; met een aanvullend onderzoek naar mogelijke verklaringen voor verschillen tussen individuele TI-curven*. (Comparison of methods for the analysis of Time-Intensity data; with reference to possible explanations for the differences between individual TI-curves.

In Dutch). Research Report, OP&P/Agricultural University of Wageningen.

van den Broek, E., Dijksterhuis, G.B., Punter, P.H., Maas, A. (1995). Effects of salivary flow and other properties of judges on Time-Intensity parameters. Unpublished paper.

van den Wollenberg, A.L. (1977). Redundancy Analysis: An alternative for canonical correlation analysis, *Psychometrika*, 42, 207-219.

van der Burg, E. (1988). *Nonlinear canonical correlation and some related techniques*. Leiden: DSWO press.

van der Burg, E., Dijksterhuis, G.B. (1989). Nonlinear Canonical Correlation Analysis of Multiway Data, In: Coppi, R., Bolasco, S. (Eds.) *Multiway Data Analysis*, 245-255, North-Holland, Elsevier Science Publishers B.V.

van der Burg, E., Dijksterhuis, G.B. (1991). An application of nonlinear redundancy analysis. In: *Statistica Applicata, Italian Journal of Applied Statistics*, 4, 65-575.

van der Burg, E., Dijksterhuis, G.B. (1993a) An application of nonlinear redundancy analysis and canonical correlation analysis. In: *Psychometric Methodology*. Proceedings of the European meeting of the Psychometric Society, Trier, Germany. In: R. Steyer, K.F. Wender, K.F. Widaman (Eds.), 74-79. Stuttgart: Gustav Fischer Verlag.

van der Burg, E., Dijksterhuis, G.B. (1993b). Nonlinear Generalized Canonical Analysis: Theory and an application from sensory research. In: *Advances in longitudinal and multivariate analysis in the behavioural sciences*. In: Oud, H., Blokland-Vogelesang, R. (Eds.), 193-203. Nijmegen: ITS.

van der Burg, E., Leeuw, J. de (1983). Non-linear canonical correlation, British *Journal of Mathematical and Statistical Psychology*, 36, 54-80.

van der Burg, E., Leeuw, J. de (1987). *Nonlinear canonical correlation analysis with k sets of variables*. RR-87-8 Enschede: University of Twente, Department of Education.

van der Burg, E., Leeuw, J. de (1988). Use of the multinomial jackknife and bootstrap in generalized nonlinear canonical correlation analysis. *Applied Stochastic Models and Data Analysis*, 4, 159-172.

van der Burg, E., Leeuw, J. de (1990). Non-linear redundancy analysis. British *Journal of Mathematical and Statistical Psychology*, 43, 217-230.

van der Burg, E., Leeuw, J. de, Dijksterhuis, G.B. (1994). OVERALS Nonlinear Canonical Correlation with *k* sets of variables. *Computational Statistics & Data Analysis*, 18, 141-163.

van der Burg , E., Leeuw, J. de, Verdegaal, R. (1988). Homogeneity analysis with k sets of variables: an alternating least squares method with optimal scaling features. *Psychometrika*, 53, 177-197.

van Trijp, H. (1992). *Sensory and Marketing research: Similarities and Differences*. Presentation at the "First Sensometrics Meeting", July 23-24, Oegstgeest, the Netherlands.

Verdegaal, R. (1986). OVERALS. UG-86-01, Leiden: University of Leiden, Department of Data Theory.

Vickers, Z.M. (1983). Magnitude estimation and category scaling of the hedonic quality of food sounds. *J. Food Sci*. 48, 1183-1186.

Vickers, Z.M. (1991). Sound Perception And Food Quality. *Journal Of Food Quality*, 1991 Feb, V14 N1, 87-96.

Wakeling, I.N., Raats, M.M., MacFie, H.J.H. (1992). A New Significance Test for Consensus in Generalized Procrustes Analysis. *Journal of Sensory Studies*, 7, 91-96.

Williams, A.A., Langron, S.P. (1984). The use of free-choice profiling for the evaluation of commercial ports. *J. Sci. Food Agric.*,35, 558-568.

Winch, R.F., Campbell, D.T. (1969). Proof? No. Evidence? Yes. The Significance of tests of Significance. *The American Sociologist*, 4, 140-143.

Yoshida, M. (1986). A microcomputer (PC 9801/MS mouse) system to record and analyze time-intensity curves of sweetness. *Chemical Senses*, 11, 105-118.

Young, F.W. (1981). Quantitative analysis of qualitative data. *Psychometrika*, 43, 247-388.

Young, F.W., Hamer R.M. (1987). *Multidimensional Scaling: history, theory, and applications*. London: Lawrence Erlbaum associates.

Young, F., Leeuw, J. de, Takane, Y. (1976) Regression with qualitative and quantitative variables: An alternating least squares method with optimal scaling features. *Psychometrika*, 41, 505-529.

Zwick, W.R., Velicer, W.F. (1986). Comparison of Five Rules for Determining the Number of Components to Retain. *Psychological Bulletin*, 99, 432-442.

ABBREVIATIONS AND ACRONYMS

ALS	Alternating Least Squares
Anova	Analysis of Variance
C	Consonance
Canals	Canonical Correlation Analysis (with optimal scaling) by Alternating Least Squares
CCA	Canonical Correlation Analysis
CA	Correspondence Analysis
Dim	Dimension
EMC	Extended Matching Coefficient
F	Fishers F (F-distribution)
FCP	Free Choice Profiling
GBP	Generalised Biplots
GCA	Generalised Canonical Correlation Analysis
Genstat	A General Statistical Program
GPA	Generalised Procrustes Analysis
GPPA	Generalised Projecting Procrustes Analysis
Homals	Homogeneity Analysis by Alternating Least Squares
Indscal	Individual Difference Scaling
IML	Instruction Matrix Language
K-sets	"more than 2" sets
Manova	Multivariate Analysis of Variance
MCA	Multiple Correspondence Analysis
MDPREF	Multidimensional Scaling of Preferences
MDS	Multidimensional Scaling
MLR	Multiple Linear Regression
MCC	Multiple Correlation Coefficient
MVA	Multivariate Analysis or Multivariate Data Analysis
NLB	Nonlinear Biplots
NPTIC	Non-centered Principal Time-Intensity Curve

OS	Optimal Scaling
Overals	Generalised Canonical Correlation Analysis (with optimal scaling) by Alternating Least Squares
PA	Procrustes Analysis
PCA	Principal Component Analysis
PCO	Principal Coordinate Analysis
PINDIS	Procrustean Individual Difference Scaling
PSA	Professional Sensory Analysis
PTIC	Principal Time-Intensity Curve
QDA	Quantitative Descriptive Analysis
RA	Redundancy Analysis
SAS	Statistical Analysis System
S-I	Sensory-Instrumental
SPSS	Statistical Package for the Social Sciences
SSQ	Sum of Squares
Statis	Structuration des Tableaux Trois Indices de la Statistique
t	Students t (t-distribution)
TI	Time-Intensity
VAF	Variance Accounted For
z-value	Standardised value (numer of standard deviations from the mean)

AUTHOR INDEX

SUBJECT INDEX

Printed and bound by CPI Group (UK) Ltd, Croydon, CR0 4YY

16/09/2026

14963413-0001